KT-412-535

Lecture Notes on
Fluid and Electrolyte
Balance

Dedicated to
Nicola, Steven and Ellen

Lecture Notes on Fluid and Electrolyte Balance

Sheila M. Willatts
MRCP FFARCS
Consultant in Anaesthesia and Intensive Care
Bristol Royal Infirmary

FOREWORD BY
Professor A. P. Adams
PhD MB BS MRCS LRCP FFARCS
Professor of Anaesthetics
in the University of London,
Guy's Hospital

SECOND EDITION

BLACKWELL SCIENTIFIC PUBLICATIONS
OXFORD LONDON EDINBURGH BOSTON
MELBOURNE PARIS BERLIN VIENNA

© 1982, 1987 by
Blackwell Scientific Publications
Editorial Offices:
Osney Mead, Oxford OX2 0EL
25 John Street, London WC1N 2BL
23 Ainslie Place, Edinburgh EH3 6AJ
3 Cambridge Center, Cambridge
  Massachusetts 02142, USA
54 University Street, Carlton
  Victoria 3053, Australia

Other Editorial Offices:
Librairie Arnette SA
2, rue Casimir-Delavigne
75006 Paris
France

Blackwell Wissenschafts-Verlag
Meinekstrasse 4
D-1000 Berlin 15
Germany

Blackwell MZV
Feldgasse 13
A-1238 Wien
Austria

All rights reserved. No part of this
publication may be reproduced, stored
in a retrieval system, or transmitted,
in any form or by any means,
electronic, mechanical, photocopying,
recording or otherwise without the
prior permission of the copyright
owner.

First published 1982
Second edition 1987
Reprinted 1992

Photocomposition by
Katerprint Typesetting Services.
Oxford
Printed in Great Britain by
Billing & Sons Ltd,
Worcester

DISTRIBUTORS

Marston Book Services Ltd
PO Box 87
Oxford OX2 0DT
(*Orders*; Tel: 0865 791155
        Fax: 0865 791927
        Telex: 837515)

USA
Blackwell Scientific Publications, Inc.
3 Cambridge Center
Cambridge, MA 02142
(*Orders*: Tel: 800 759-6102)

Canada
Times Mirror Professional Publishing, Ltd
5240 Finch Avenue East
Scarborough, Ontario M1S 5A2
(*Orders*: Tel: 416 298-1588)

Australia
Blackwell Scientific Publications
(Australia) Pty Ltd
54 University Street
Carlton, Victoria 3053
(*Orders*: Tel: 03 347-0300)

British Library
Cataloguing in Publication Data

Willatts, Sheila M.
  Lecture notes on fluid and electrolyte
  balance.—2 ed.
  1. Acid base equilibrum   2. Body
  fluids   3.   Electrolyte metabolism
  I. Title
  612'.01522   QP90.5

  ISBN 0-632-01714-7

Contents

Foreword

Concoctions of fluids and substances of every conceivable kind have been administered by one route or another for centuries as a treatment for disease, and as often as not such medication appears to us nowadays as being worse for the patient than the original malady. Fashions in concept and treatment tend to change almost every decade and so both old as well as new substances and therapies require constant reappraisal.

Although the importance of fluid balance and electrolyte composition is germane to the proper function of every organ, tissue and cell in the body, it is a subject which most doctors find difficult to understand, to remember or to apply except in the most superficial way.

It is vital that a short but clear and unambiguous text should appear on this subject as it affects practically every discipline in the practice of medicine. This is no easy task because we all tend to be only interested in the aspects we like of the subject: we are not all necessarily involved in every detail but we need to become more familiar with the subject in its widest sense.

This book has been written to enable the commonsense and practical approach to complement the scientific data. Sheila Willatts has an international reputation as a teacher and lecturer; her expertise as an anaesthetist and her interest in acute intensive care medicine well befits her authorship of this intricate subject. This second edition has been extensively revised and reflects the real increase in information which is available today.

Anthony P. Adams
Professor of Anaesthetics
in the University of London,
United Medical & Dental Schools,
Guy's Hospital, London

Preface to Second Edition

Since 1982 there has been a great deal of work and evaluation in the field of fluid and electrolyte therapy. In particular the role of sodium and other electrolytes in the genesis of hypertension has received intense scrutiny. Our understanding of the role of calcium in tissue metabolism has increased and calcium channel blocking drugs are in widespread use.

In this Second Edition normal acid base balance has been separated from disturbances thereof. The metabolic response to trauma is now incorporated into the completely rewritten chapter on nutrition and Fluid Balance in Special Circumstances is expanded. Most chapters have been updated in the light of current thinking and some discussion introduced, in a way which I hope does not detract from the clarity. Once again I am indebted to my colleagues at Bristol Royal Infirmary and Frenchay Hospital for their forebearance.

Sheila M. Willatts
1986

Preface to First Edition

Fluid and electrolyte balance is a subject about which few wax lyrical. Cynics may state that our understanding in this difficult area has improved little over the last two decades. However, considerable advances have been made in our understanding of calcium and phosphate metabolism and technological advances now permit continuous recording of plasma electrolytes by intravascular ion selective electrodes. More practically, parenteral nutrition, for example, can be undertaken scientifically and dialysis allows manipulation of total body water and electrolytes.

This book is not for the specialist, neither does it deal with paediatric fluid balance. The first part outlines normal water, electrolyte and acid base balance and is concluded by tables of normal values. Part II considers disturbances from an essentially practical point of view concluding with a chapter on setting up infusion lines. The text is aimed at senior students and junior medical staff. Although much of the book is dogmatic some discussion is included on more recent aspects.

I am indebted to many colleagues at Frenchay Hospital and Bristol Royal Infirmary for advice, for reading and commenting on the text and in particular to my anaesthetic colleague Dr Frank Walters for his numerous helpful suggestions. I should like to thank the secretaries at Frenchay Hospital for initial typing and Peter Cox for providing illustrations. Special thanks are due to Norma Bryden who typed the final manuscript so efficiently and to Blackwell Scientific Publications for their help and encouragement during the writing of this book.

I hope that those who read it will subsequently take a greater interest in the management of fluid therapy in their patients.

The book is dedicated to my three children without whose help it would have been finished in half the time.

Sheila M. Willatts
1982

Glossary of Terms

The International System of Units (SI, Système International) is in widespread use in Great Britain and is adopted here. Those values which are relevant in this book include:

 units of mass — kilogram (kg)
 amount of a substance — mole (mol)
 energy — joule (j)
 pressure — pascal (Pa), kilopascal (kPa)

The preferred volume is a litre (l). Chemical analyses therefore report mmol l^{-1} or μmol l^{-1}. Plasma proteins are still reported in g l^{-1} and Hb as g per decilitre (g dl^{-1}).

ABBREVIATIONS

AA	amino acid	CCF	congestive cardiac failure
ACD	acid citrate dextrose		
ACTH	adrenocorticotropic hormone	Cl	chloride
		cm	centimetre
ADH	antidiuretic hormone	CO	cardiac output
ANP	atrial natriuretic peptide	CO_2	carbon dioxide
		COP	colloid osmotic pressure
ARDS	adult respiratory distress syndrome		
		CPD	citrate phosphate dextrose
ARF	acute renal failure		
ATP	adenosine triphosphate	Cr	creatinine
		CSF	cerebrospinal fluid
ATPase	adenosine triphosphatase	CVP	central venous pressure
BP	blood pressure	CVS	cardiovascular system
Ca	calcium	DPG	diphosphoglycerate

DVT	deep vein thrombosis	kg	kilogram
ECF	extracellular fluid	l	litre
ECG	electrocardiogram	LAP	left atrial pressure
EEG	electroence-	LV	left ventricle
	phalogram	Mg	magnesium
FFA	free fatty acids	ml	millilitre
g	gram	mm	millimetre
GFR	glomerular filtration	MW	molecular weight
	rate	N_2	nitrogen
H	hydrogen	Na	sodium
HAS	human albumin solu-	NH_3	ammonia
	tion	NH_4	ammonium
Hb	haemoglobin	O_2	oxygen
HCO_3	bicarbonate	OH	hydroxyl
HD	haemodialysis	PCWP	pulmonary capillary
Hg	mercury		wedge pressure
HPPF	human plasma	PD	peritoneal dialysis
	protein fraction	PO_4	phosphate
HPT	hyperparathyroidism	PTH	parathormone
ICF	intracellular fluid	RA	right atrium
ISF	interstitial fluid	RNA	ribonucleic acid
iv	intravenous	RV	right ventricle
JVP	jugular venous	SA	Siggaard–Andersen
	pressure	TBW	total body water
K	potassium		

PART I

Chapter 1
Normal Water Balance and Body Fluid Compartments

Water, water everywhere . . . nor any drop to drink.

The Ancient Mariner
SAMUEL TAYLOR COLERIDGE

Water is indeed to be found everywhere in the human body. Without it survival is limited to a few days in the adult whereas total food deprivation is tolerated for at least a month. This fact is well known to surgeons and anaesthetists presented with patients who have advanced carcinoma of the oesophagus and severe dysphagia. There is no evidence for evolutionary adaptation to water deprivation in man. However, the movement of vertebrates from a salt water to a fresh water environment lead to adaptive mechanisms in the distal tubule to protect the body from fatal dilution.

TOTAL BODY WATER

Total body water (TBW) constitutes 60% of body weight in the male and 52% of body weight in the female (due to the higher fat content) with one third of this in the extracellular space and two thirds intracellular.

Distribution of body water

A 70 kg man will have:
  TBW 42 litres.
  Extracellular fluid (ECF) 14 litres (20% of body weight in males and females).
  Intracellular fluid (ICF) 28 litres.
  Plasma volume 3.5 litres (25% of ECF).

Studies of body composition show physiological changes with ageing. TBW is reduced with the greatest reduction occurring in ECF.

Body water is passively distributed between ECF and ICF according to the osmolar content of each compartment.

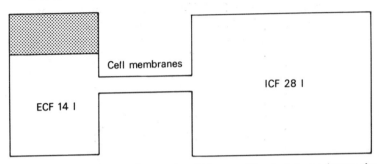

Fig. 1.1 Distribution of body water in a 70 kg man. The shaded area denotes the intravascular compartment.

As water is highly polar it is an ideal biological solvent for polar compounds, allowing stable dissociation of ions for intermediary metabolism. ECF is the transport medium for nutrients and waste products of metabolism whereas ICF is required for their structure, organization and function. The monovalent ions, sodium and potassium are hydrated within the cell; each millimole binding 7 ml water. Intracellular water is not uniformly distributed. The highest concentration is present in metabolically active cells such as muscle, liver and kidney whilst the lowest concentration is found in inactive structures such as bone.

Measurement of fluid compartments

Total body water
ECF volume
  Blood volume
  Interstitial water
Intracellular volume
Transcellular fluid

Measurement of total body water

  Deuterium oxide D_2O.
  Tritium oxide
  Antipyrine

Deuterium oxide is a stable compound which distributes itself as water throughout the body. It is not an isotope. Antipyrine is rarely used as it is not readily available in this country. Its main value is in circumstances when a scintillation counter is not available. Tritium oxide ($^{3}H_{2}O$) is an isotope of water and as such distributes throughout all compartments in the same way as water. The basis of isotope dilution measurement lies in the injection of an isotope which will distribute completely throughout the compartment to be measured within a finite time, after which its concentration (Cd) can be easily measured. During the measurement the subject must be in a stable metabolic state and allowance must be made for any excretion or metabolism which may occur during the equilibration period. If a given concentration of an isotope (Ci) is introduced into a fluid compartment in a volume (Vi) the distribution space (Vd) may be calculated at equilibrium thus:

$$Vd = \frac{Ci \times Vi}{Cd}$$

Tritium oxide is the isotope of choice for TBW measurement as mixing is good after 3–4 hours, metabolism is insignificant and urine losses are very small. It is a weak beta-emitter with a biological half life of 10 days which makes it suitable for measurement in humans, although the radioactive (physical) half life is 12 years. 0.5 millicuries of $^{3}H_{2}$ are injected and 4 hours allowed for normal equilibration although 6–8 hours may be required for the obese patient or those with ascites. The reproducibility of this method is usual ±2%.

Measurement of ECF volume

1 **Crystalloids:** inulin, mannitol or sucrose. These are large molecules and therefore do not penetrate the entire ECF. Failure to penetrate areas such as the digestive tract and cerebrospinal fluid (CSF) will lead to an underestimation of the ECF volume.

2 **Ionic substances:** isotopes of chloride, bromide, sodium and sulphate. These are smaller molecules but although they penetrate the entire ECF they may also penetrate cells and therefore overestimate the ECF volume.

Studies of the disappearance of injected isotopes from the plasma suggest two rates of transfer:

(a) Transfer to a rapidly equilibrating pool in dynamic equilibrium with plasma (the functional ECF). This includes 25% of dense connective tissue and 10% of bone water. The volume of this compartment is 8.4 litres (l).

(b) Transfer to a slowly equilibrating pool over 24 hours. This pool includes the remaining connective tissue and bone.

MEASUREMENT OF BLOOD VOLUME

Measurement of plasma volume.   Three methods are available:

1   Evans blue dye is used to bind to plasma albumin.
2   Radioiodine is used to label serum albumin (RISA).
3   Labelled macroglobulin may be injected.

The disadvantage of the RISA method is the escape of 7–10% of the iodinated albumin per hour into interstitial fluid thereby causing an overestimate of plasma volume which becomes even greater in burns, trauma and ascites. This problem can be overcome to some extent by reducing the equilibration period.

Measurement of red cell volume.

1   This may be calculated from the plasma volume and haematocrit.
2   The volume of distribution of reinjected red cells labelled with radioactive chromium (^{51}Cr) may be calculated.

MEASUREMENT OF INTERSTITIAL WATER

Interstitial water is all that ECF which does not constitute intravascular fluid. It is calculated by subtracting plasma volume from ECF volume.

Measurement of intracellular fluid volume

The volume of ICF cannot be measured directly. It is therefore calculated as follows:

$$ICF = TBW - ECF.$$

Such a calculated value will include the errors from both of the component measurements.

There is good correlation between ICF and total exchangeable potassium. Since potassium is the main intracellular cation its depletion leads to movement of water from the cells into ECF. Thus ICF volume depends on ECF osmolality such that a fall in ECF osmolality increases ICF volume since water enters the cells by osmosis (see below). ICF volume also depends on muscle mass and decreases with age.

TRANSCELLULAR FLUID

This is that part of the ECF which is formed by the transport activity of cells and includes secretions into the lumen of the gastrointestinal tract and the CSF.

The water content of body fluid compartments has been outlined so far without considering the solute content. Table 1.1 shows the mean ionic content of the body fluid compartments in mmol 1^{-1}. The anion gap is explained in Chapter 13 but is included here for the sake of completeness.

Table 1.1 Mean ionic composition of body fluid compartments (mmol 1^{-1}).

Substance	Plasma	Interstitial fluid	Intracellular fluid
Sodium	141	144	10
Potassium	3.7	3.8	156
Chloride	102	115	3
Bicarbonate	28	30	10
Anion gap	15	–	–
Calcium	2.4	–	–
Magnesium	0.8	–	11
Phosphate	1.1		31
Protein	16	10	55

It is clear that there are striking differences in concentration of ions between compartments. This is further illustrated in Fig. 1.2. The most important factor maintaining separation of these compartments is the cell membrane: a complex structure consisting of lipids and proteins.

Despite the differences in ionic concentration between compartments, the law of electrical neutrality states that the sum of the

negative charges of anions must be equal to the sum of positive
charges of cations in any compartment. The ways in which regulation of fluid between compartments takes place will now be
considered.

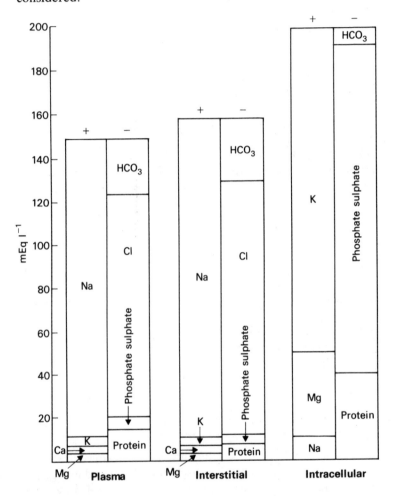

Fig. 1.2. Plasma, interstitial and intracellular electrolyte concentrations.

PHYSIOLOGICAL REGULATION OF
BODY FLUID COMPARTMENTS

The following factors are important:
1 Osmosis.
2 Diffusion.
3 Gibbs Donnan equilibrium.
4 Starling's forces.
5 Sodium pump.

Osmosis

This is the movement of solvent molecules across a selectively permeable membrane from a region of low solute concentration to a region of high solute concentration the membrane being impermeable to solute molecules.

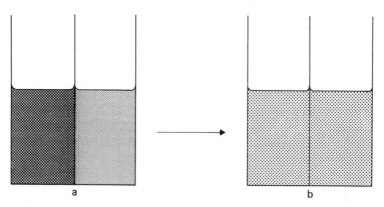

Fig 1.3 Solutions of different solute concentrations separated by a permeable membrane. In (a), the solutions are separated by a permeable membrane and equilibrium takes place readily as shown in (b), where equal concentrations occur on either side of the membrane.

Biological membranes are selectively permeable so that when solutions of different solute concentrations are separated solute molecules cannot equilibrate. Instead water passes from the region of low solute concentration to a region of high solute concentration.

All cell membranes and capillary walls within the body are freely permeable to water. Hence all fluid compartments within the body have the same effective osmolality as plasma. Some solutes can permeate across cell membranes and achieve their own equilibrium (see below). Following an acute change in the composition of a fluid compartment equilibration rapidly occurs with water moving down its concentration gradient to the solution of higher osmolality.

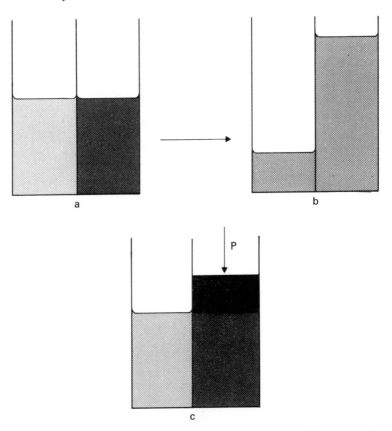

Fig. 1.4 Solutions of different solute concentration are separated by a selectively permeable membrane, (a). The selectively permeable membrane separates two solutions of different solute concentrations; (b), water crosses the membrane until the solute concentrations are equal; (c), application of a pressure P which is the osmotic pressure will stop the movement of water (solvent) molecules across the membrane.

Osmolality and osmolarity

The osmotic activity of a substance in solution depends only on the *number* of discrete particles dissolved and not on their weight or valency. Hence a fully ionized substance will have double the osmotic effect of the same molar concentration of a substance in the unionized state.

OSMOLE

One osmole of a substance in ideal solution depresses the freezing point of a solvent, in this case water by 1.86 °C. One milliosmole (mosmol) is one thousandth of an osmole.

One mole of any substance contains the same number of molecules, (Avogadro's number 6.061×10^{23}) and one mole = 1000 millimoles (mmol). A mole is the molecular weight (MW) of a substance in grams (g). A mole of an unionized substance is equal to an osmole but one mole of completely ionized sodium chloride equals two osmoles. In fact body fluids are not ideal solutions owing to interactions between ions.

OSMOLALITY

The osmolality of a solution is the number of osmoles of solute per kilogram of solvent. Clinically, this is more usefully expressed as milliosmoles per kg (mosmol kg^{-1}).

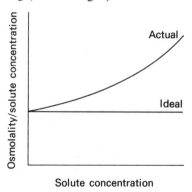

Fig. 1.5 A non ideal solution showing an apparent rise in osmolality with increasing solute concentration.

In an ideal solution the ratio of osmolality to molal concentration remains constant as the molar concentration changes. However, in practice the ratio increases as the solute concentration rises because of interaction between solute and solvent.

The osmolarity of a solution is the number of osmoles of solute per litre of _solution._ Hence it is temperature dependent.

This value may be calculated from plasma concentrations of electrolytes, glucose and urea. At least 17 formulae exist for calculation of osmolarity; one of the most acceptable is

$$2(\text{sodium} + \text{potassium}) + \text{glucose} + \text{urea}.$$

all values being in mmol 1^{-1}.

Values obtained in this way may be greater than measured osmolality as sodium chloride is only about 93% dissociated in plasma. Measured osmolality may exceed the calculated value in the presence of unmeasured osmotically active particles, for example mannitol and ethanol.

Methods of measurement of osmolality

1 Freezing point osmometers. These instruments depend on the principle that depression of freezing point of a solution is proportional to its total osmolar content.

2 Vapour pressure osmometers. The water vapour pressure of an aqueous solution varies with its osmolality. These instruments make use of this principle and are accurate to within ± 3 mosmol kg^{-1}. They can operate on very small samples (8 μl).

Normal plasma osmolality is 280–295 mosmol kg^{-1}.

Urine osmolality varies from 30–1400 mosmol kg^{-1}.

An isotonic solution is one with the same osmolality as plasma; thus red cells suspended in it do not change their volume. A hypotonic solution is one in which the osmolality is less than that of plasma. A hypertonic solution is one with an osmolality greater than that of plasma. If a measurement of osmolality is not available the calculated plasma osmolarity may be used. The difference between these two values is the osmolar gap and is usually 0–24 mosmol kg^{-1}.

Although osmolality measurement is requested most often in cases of hyponatraemia, it is of limited value in this circumstance but does have two specific uses; to determine whether plasma water content differs from normal and to screen for the presence of low molecular weight substances in the blood. Osmolality gives an indication of the specific activity of water in a solution. Tonicity can be equated with effective osmolality and is therefore a measure of water movement across a semipermeable membrane. Calculation of tonicity by adding non-permeate solutes allows rapid estimation of the state of hydration in hyperglycaemia. Except in rare circumstances where plasma water content is markedly reduced or large amounts of mannitol are present in the circulation a deviation from normal of calculated osmolarity indicates an abnormality of intracellular as well as extracellular hydration (i.e. an abnormality in tonicity).

Osmolar gap.   The osmolar gap may be raised in:
1   Ethanol intoxication. Ethanol depresses the freezing point of plasma water. Mannitol, methanol, ethylene glycol and other toxins with MW less than 150 produce the same effect.
2   Multiple myeloma and other hyperproteinaemic states.
3   Hyperlipoproteinaemia.
In these last two conditions the percentage of plasma solids is raised from the normal 7% so that although the concentration of electrolytes in plasma water is normal (that is to say, the measured osmolality is normal) the percentage of the plasma that is really water is reduced. Thus there is a reduction in calculated osmolarity. Serum sodium is low but not usually less than 120 mmol 1^{-1}. There is also an increase in osmolar gap in ketoacidosis, chronic renal failure and shock with lactic acidosis. These conditions are also associated with an increased anion gap.

Indications for measurement of osmolality.
1   Screening for renal disease, tests of urine concentrating ability.
2   Following the course of intrinsic renal disease.
3   Calculation of osmolar and free water clearance.
4   Differential diagnosis of polyuria.
5   Differential diagnosis of oliguria of sudden onset.
6   Diagnosis of inappropriate ADH secretion.

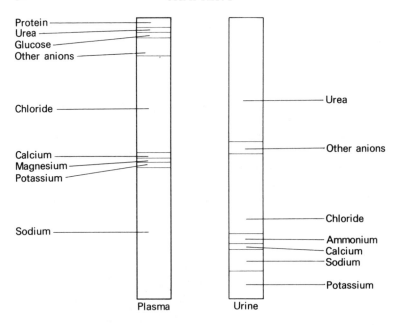

Fig. 1.6 Constituents of plasma and urine.

Urine constituents vary widely especially during illness so that calculations for urine osmolarity are very prone to error. However, in the absence of a measured osmolality the measured specific gravity, which is the weight of a solution compared with the weight of water, is of some help. In the absence of hyperglycaemia, proteinuria or excretion of radiological contrast media the relationship shown in Table 1.2 applies. Problems arise if patients have been treated with diuretics, especially osmotic agents, which cause increased Na and water excretion with a relatively high urine SG even in the presence of volume and Na depletion.

Table 1.2 Relationship between urine specific gravity and osmolality.

Specific gravity	Osmolality
1.010	350
1.020	700
1.030	1050

A recent developed reagent strip provides a guide to urine specific gravity and osmolality with reasonable accuracy.

ECF OSMOLALITY

The main determinant of ECF osmolality is its sodium concentration. If ECF sodium concentration and hence osmolality rises water will be withdrawn from cells to re-establish equal osmolality (isotonicity). Conversely acute hyponatraemia will increase intracellular water regardless of the level of total body water. Sodium is unique as a solute in that changes in its concentration are more often due to changes in the volume of its solvent (water) than in total sodium stores.

ICF OSMOLALITY

The main determinant of ICF osmotic pressure is the intracellular potassium concentration. Some intracellular ions may be osmotically inactive, for example, 30% of intracellular magnesium is bound to lipoproteins, ribonucleic acid (RNA) and adenosine triphosphate (ATP).

It is not clear whether intracellular K or extracellular Na is the prime controller of body osmolality.

COLLOID OSMOTIC PRESSURE

Measurement of plasma osmolality is almost entirely a measure of crystalloid osmolality. Only about 0.5% of the total plasma osmolality is attributable to colloid. Colloids are large protein molecules with a molecular weight greater than 20–30 000. This small percentage, however, is very important since capillaries are only minimally permeable to colloids which therefore have an important effect of 'holding' water within the plasma compartment and maintaining water distribution between body fluid compartments. A fall in colloid osmotic pressure is an important aetiological factor in the development of pulmonary oedema. Normal colloid osmotic pressure (COP) is 20–25 mmHg being greater in the ambulant than in the supine patient. This COP is largely attributable to plasma proteins although the correlation between plasma albumin and COP under stable conditions is unreliable.

Recently developed machines for colloid osmotic pressure measurement (oncometers) utilize semipermeable membranes and electronic pressure transducers to measure COP on small samples of fluid.

Diffusion

Simple diffusion permits passage of a substance along a concentration gradient. Cell membranes have a high lipid content hence lipophilic and unionized substances will cross membranes at a faster rate than hydrophilic, polar or ionized substances. Diffusion is facilitated by pores in cell membranes of 0.7 nanometres diameter. The rate of diffusion of a substance does not depend on its atomic weight but its hydrated radius, that is to say its size when hydrated with water molecules.

Gibbs Donnan equilibrium

ICF contains more anionic protein molecules than does interstitial fluid. This results in an increase in diffusible cations (sodium, potassium) and a decrease in diffusible anion (chloride) and means that the total number of diffusible ions is greater in ICF (see Fig. 1.2).

Starling's forces

In an adult the capillary bed provides 6300 square metres of filtering surface. Plasma protein osmotic pressure is about 25 mmHg whereas blood pressure is 35 mmHg at the arterial end of the capillary and 15 mmHg at the venous end. This results in diffusion of water and diffusible ions out of the capillary and into the interstitial fluid at the arterial end of the capillary and reabsorption of about 90% of this at the venous end.

Total flow of water from the interstitial fluid is 20 l daily in an adult. Normal interstitial fluid pressure is about 7 mmHg subatmospheric. Normally lymphatics drain excess fluid from the interstium and lymph flow can increase dramatically if the interstitial fluid is rapidly expanded.

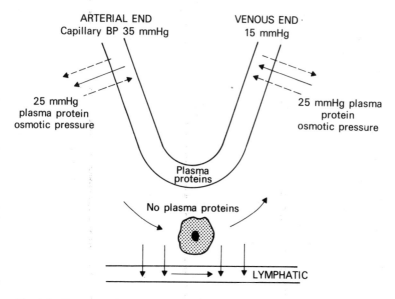

Fig. 1.7 Transport of water in the capillary circulation.

Sodium pump

The non-diffusible intracellular proteins tend to draw water into cells. Sodium (Na) tends to enter the cell by diffusion from a high ECF to a low ICF concentration. This tendency is offset by the sodium pump which is located in the cell membrane and actively extrudes Na from the cell into ECF. The sodium pump depends on a supply of ATP, although this inhibition is partly antagonized by K. It is very temperature sensitive and is inhibited by cardiac glycosides, a similar, linked mechanism exists for transport of potassium (K) ions in the opposite direction.

The sick cell syndrome occurs in a variety of situations such as hypoxia and septicaemia when the sodium pump breaks down, Na enters the cell and K is lost from the cell into plasma. The ions move down their concentration gradients. Plasma Na falls and plasma K usually rises.

As a result of unequal distribution of ions across membranes a potential difference of 10-100 millivolts is set up across the cell membrane with the inside negative to the outside. This potential difference is closely related to the difference in K concentration across the membrane and may be described by the Nernst Planck equation.

This assumes that at the resting potential of biological membranes permeability to sodium and to chloride is zero.

$$V = \frac{RT}{F}\log_e\frac{Ko}{Ki}$$

V = Potential difference
R = The gas constant
F = Faraday's constant
T = Temperature
Ko = Concentration of potassium in the ECF
Ki = Concentration of potassium in the ICF

Total body water content is finely regulated. Any attempt at parenteral repletion of body fluids requires a good working knowledge of the basic normal physiological control of water balance.

WATER BALANCE

To replace insensible losses and excrete solute products of metabolism, adults require approximately 1500 ml water m^{-2} daily.

Water intake

1   Oral fluids
2   The water content of food. Melon and citrus fruits contain about 90% water.
3   Metabolism. Oxidation of food to carbon dioxide and water provides 300–500 ml of water daily (120 ml per 1000 calories). Water is absorbed in response to the osmotic pressure difference between plasma and intestinal contents. At any one time about 1.5% of body water lies within the gastrointestinal tract with food and electrolytes. Most gastrointestinal contents are isotonic although saliva is hypotonic.

Despite processing ingested water is far from pure. Acid rain accounts for pollution by aluminium, manganese, zinc, nickel, lead and cadmium. Some of these reach concentrations in lakes which are toxic to animal life, for example the high mercury content in fish from acidified waters in the US and Scandinavia.

Aluminium toxicity is recognized in patients with renal failure, in the aetiology of amyotrophic lateral sclerosis (in Guam) and is linked to Alzheimer's disease.

Water loss

1   Urine. Urine volume is very variable but on average amounts to 1500 ml per day. A minimum volume of 500 ml in an adult is necessary to eliminate the normal solute load for excretion. The elderly are less efficient at excreting a water load during the day so that nocturia and sleep disturbance are more common than in younger patients.

2   Insensible loss. This amounts to 0.5 ml kg^{-1} $hour^{-1}$ at 37 °C (840 ml in 24 hours in a 70 kg man). Evaporation occurs from the skin and water vapour is lost through expired air. Evaporative water loss from the respiratory tract increases by 15% during surgery.

3   Faeces 100–200 ml daily.

This simple outline of water balance takes no account of pathological processes which will be dealt with in subsequent chapters. Where physiological changes such as ageing reduce TBW, especially ECF, factors which control water balance may have more profound effects. Individual factors vary greatly from day to day. Muscular work and fever produce a marked increase in water loss. Sweating occurs from sweat glands controlled by the sympathetic nervous system from the hypothalamic heat regulating centre. A hot environment and muscular activity increase sweat production. In hot climates up to 3 litres of water per hour may be lost as sweat. Very obese patients have markedly increased insensible water loss, largely as sweat which may amount to 3 l per day. Salicylate poisoning, thyrotoxicosis and other factors which increase metabolism will also increase water loss.

Evaporation of sweat will increase heat loss (latent heat of vaporization).

Two simple measurements may help in elucidating the water balance problems in an adult:

1   Weigh the patient. Acute changes in weight are likely to be due to changes in water balance. Starvation is considered in later chapters.

2   Check that the urine volume is greater than 1 litre in 24 hours.

Regulation of water balance

Happily for humans osmotic homeostasis can be maintained over a wide range of water intake provided the following are normal:
1 Renal function.
2 Renal blood supply.
3 Hormonal activity.
The fact that many patients come to no harm and some even do well following very varied, often inappropriate intravenous fluids prescribed by well intentioned medical staff probably reflects the normal ability of the kidney to vary the solute excretion to maintain normal plasma osmolality. Daily water turnover depends on osmolar production and renal concentrating ability. Fine regulation of water balance keeps plasma osmolality between 280 and 295 mosmol kg^{-1}. Under normal circumstances daily water requirements are 30 ml kg^{-1}. The minimum water intake is that which is required to replace loss from all body sources and the maximum intake is that which can be excreted by the kidney.
Minimum urine output =

$$\frac{\text{osmolar load (mosmol per day)}}{\text{renal concentrating ability (mosmol } 1^{-1})}$$

Normally this is $\dfrac{600}{1200}$ = 0.5 litres per day

In health the possible range of water intake is very wide before any disturbance occurs, but in the ill, hypercatabolic patient the range may be small owing to an increased solute load, excess losses, impaired renal function, falling cardiac output and the hormonal imbalance of the stress response. In such a patient the osmolar load may rise to 1000 mosmol per day and concentrating ability may be reduced to 800 mosmol 1^{-1}.

Minimal urine output then becomes $\dfrac{1000}{800}$ = 1.25 l per day

In addition to this increased minimum urine output, fever will increase insensible water loss.

Many hormones influence salt and water absorption from the intestine. Corticosteroids and aldosterone in particular are important where salt depletion or overload may occur. It is possible that some gastrointestinal hormones liberated locally limit the rate of absorption of sodium and water.

Variations in water intake or loss produce a small plasma osmotic change which activates regulatory mechanisms via the hypothalamus affecting thirst and urine concentration.

Thirst

Thirst is more important than urine concentration in preventing dehydration due to water deprivation or due to the hyperosmolality of an increased solute load. Pronounced overactivity of the renin angiotensin system acting via angiotensin II can induce increased sensitization of the cerebral sensors exerting osmotic regulation of water intake and antidiuretic hormone release.

The most usual circumstance is that water intake is in excess of needs, and this excess water is excreted by the kidney.

ROLE OF THE KIDNEY IN WATER EXCRETION

The renal glomeruli produce 120 ml per minute of filtrate from plasma (180 l over 24 hours). In the case of water 99% of this will be reabsorbed as the filtrate passes through the tubule thereby producing about 1.5 l of urine per day. Normally there is a wide range of urine osmolality such that the same solute load can be excreted in 500 ml of urine with an osmolality of 1400 mosmol kg^{-1} or in 23.3 l of urine with an osmolality of 30 mosmol kg^{-1}. Table 1.3 illustrates the wide range of urine volume possible under different clinical circumstances.

Table 1.3 Range of urine volume.

	Osmolar production (mosmol daily)	Maximum urine osmolality (mosmol kg^{-1})	Minimum urine osmolality (mosmol kg^{-1})	Maximum urine volume (1)	Minimum urine volume (1)
Healthy	600	1200	30	20	0.5
Hypercatabolic	1200	1200	30	40	1
Postoperative on 2 litres of 5% dextrose daily	200	600	200	1	0.3

Countercurrent multiplier

A gradient of increasing osmolality exists as urine progresses along the tubule more deeply into the medulla allowing concentration to take place (multiplier).

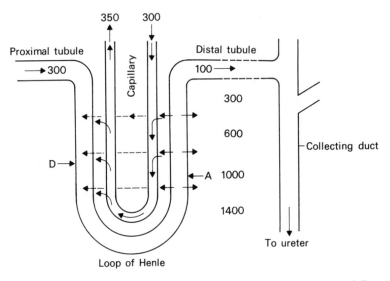

Fig. 1.8. Countercurrent system for production of urine. Numbers = osmolality; D = descending limb; A = ascending limb.

25% of filtered Na is reabsorbed between the end of the proximal tubule and the beginning of the distal tubule whereas less than 15% of filtered water is reabsorbed in this region. Therefore more salt is reabsorbed than water and dilute fluid is present at the end of the distal tubule. The rate of salt reabsorption in the loop of Henle varies with the sodium load presented to it by the proximal tubule. Recently it has been shown that Na reabsorption in the ascending limb of the loop of Henle is due mainly to active transport of chloride (Cl) ions. This renders the lumen of the tubule electrically positive (3–9 mV) with respect to the peritubular fluid. Therefore Na may diffuse out *passively* along the electrical and chemical gradient. This segment of the loop of Henle is virtually impermeable to water. Na diffuses passively into the descending limb and is carried round again to the ascending limb where the process is repeated. The blood in the accompanying capillary also takes part in this exchange. Na concentration increases as blood flows through the hypertonic medulla and it decreases again when the capillary leaves the area of high osmolality.

The countercurrent mechanism depends on the hairpin structure of the loops of Henle and the high concentration of Na, Cl and urea in the depths of the medulla. The collecting ducts act as a source of urea to add to the interstitial hypertonicity. This mechanism depends on two factors:

1 Solute impermeability of the descending limb of the loop of Henle.

2 Permeability of the ascending limb to Na, Cl and urea but NOT to water.

Fluid in the descending limb therefore becomes concentrated by removal of water, and fluid in the ascending limb is diluted by net loss of salt in excess of the net gain of urea by passive diffusion. The composition of the interstitial fluid (a mixture of hyperosmotic concentrations of salt AND urea) is very important for this process. Fluid in the descending limb of the loop of Henle contains only salt as the osmotically active component (Figs. 1.9, 1.12).

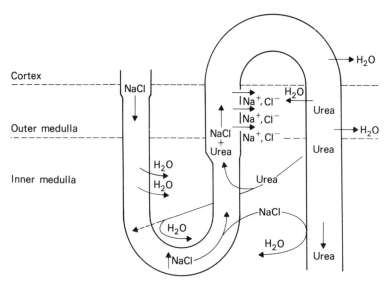

Fig. 1.9 Countercurrent concentrating mechanism.

Fluid entering the lower ascending limb contains sodium chloride at higher concentration than the surrounding fluid which has the same osmotic pressure but due to *urea + salt*. Therefore in this segment there is net outward diffusion of NaCl and inward diffusion of urea down their respective concentration gradients. More salt must be lost than urea gained because fluid at the distal end of the loop is hypotonic to plasma. The general view is that transport of NaCl out of this thin ascending limb is the active step in countercurrent multiplication.

ROLE OF UREA

When urea is the principal constituent of urine the volume of urine required to excrete a given solute load is reduced. Urea clearance is always less than glomerular filtration rate (GFR) and is independent of the plasma urea over a wide range of concentration. Urea clearance depends on urine flow (fractional water reabsorption). The mechanism for this is the recycling of urea from the medullary collecting ducts into the loops of Henle and is made possible by the almost continuous reabsorption of water along the nephron. Urea transport is passive and as water is reabsorbed urea becomes more concentrated and this is maximal in the medullary collecting ducts. This effect is augmented by an antidiuresis. Even at high antidiuretic hormone (ADH) levels the permeability of collecting ducts and distal tubules to urea is low, therefore increased water reabsorption very effectively progressively raises the urea concentration of fluid entering the medullary collecting ducts. More urea is reabsorbed in the medulla at low urine flows when hypertonic urine is formed than at high urine flows. The reabsorbed urea diffuses into the medullary interstitium and vasa recta until its concentration is nearly the same at any given level as in the collecting duct.

Urea constitutes 40–50% of the total solute concentration of the medullary interstitial fluid during antidiuresis and 10% during a water diuresis.

A high urea in glomerular filtrate therefore promotes increased urinary concentration of non-urea solutes. The urea induced passive concentration of salt in the descending limb and increased outflow of salt from the ascending limb into the interstitium augments a single passive effect in this segment. Therefore, a rise in medullary salt concentration is achieved leading to increased water reabsorption from collecting ducts and hence an increase in non-urea solutes in the urine. The concentrations of Na and urea in the interstitium are kept high by the slow blood flow in the vasa recta.

ADH may contribute to concentration of urine not only by enhancing the permeability of the distal tubule but also by enhancing the transport of salt and urea into the medullary interstitium to increase the osmotic gradient.

An efficient countercurrent system depends on:

1   A good glomerular filtration rate. A reduction in GFR of 25% will reduce concentrating ability significantly because of inadequate delivery of Na and urea to the tubule. Sufficient sodium chloride must be available to the ascending limb active transport system to maintain the hypertonicity in the medulla.

2   If the solute load is increased this has the effect of washing out medullary osmoles and reduces the time the multiplier has to work. In these circumstances both concentrating and diluting ability are affected. Flow rate in the collecting ducts is increased and diluting ability reduced. This is illustrated in Fig. 1.10. Osmotic diuretics effectively increase the solute load.

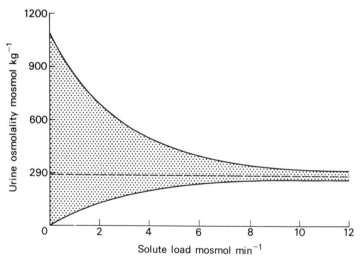

Fig. 1.10   The effect of solute load on urine osmolality.

Osmotic diuresis

A solute such as mannitol that is not reabsorbed in the proximal tubule exerts a considerable osmotic effect. As its concentration in the tubule rises large volumes of isotonic fluid are presented to the countercurrent mechanism. More fluid passes through the loops of Henle, the hyperosmolality of the medullary pyramids is reduced and less water is subsequently reabsorbed from the collecting ducts.

Free water

Daily fluid intake often exceeds requirements so that the kidney is usually required to excrete the excess water. The kidney can excrete water in excess of that required to render the urine isotonic as we have seen above provided the secretion of antidiuretic hormone can be suppressed. Such excess water excretion is termed free water. If excess ADH is secreted, as in the metabolic response to trauma (see Chapter 12) which overrides normal osmoreceptor regulation, then any free water will be reabsorbed in the distal tubule and collecting duct. In these circumstances the only route of free water excretion is through the lungs and skin.

WATER DEPRIVATION

Water deprivation or increased solute intake produces thirst.

Thirst

Thirst sensitive areas exist in the ventromedial and anterior areas of the hypothalamus. This area is closely associated with nuclei regulating ADH secretion. Therefore if water is withheld, hypertonicity is detected by osmoreceptors and neural signals sent to the thirst centre (Fig. 1.11). Thus:
1   A sensation of thirst occurs.
2   ADH is secreted.
3   Food intake is inhibited.
Angiotensin II which is increased in ECF depletion is a very potent stimulus for thirst if injected into the brain. Taste and other oropharyngeal sensations are also important in thirst regulation.

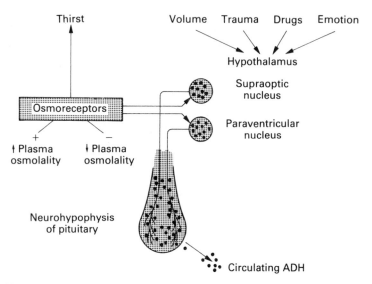

Fig. 1.11   Relationship between plasma osmolality, thirst and ADH secretion.

Thirst increases water ingestion in the conscious patient. This water moves freely across the mucosa of the small and large intestine and to a lesser extent across the gastric muscosa. If a large amount of hypotonic fluid is ingested in the absence of water depletion, water diuresis occurs after 15 minutes. This represents the time required for absorption of the water and inhibition of ADH secretion. The diuresis is maximal at about 40 minutes after ingestion. Maximum urine flow during a water diuresis is about 16 ml min^{-1}.

Osmoreceptors

The osmoreceptor area in the hypothalamus can detect a 1% change in body osmolality which in an adult is roughly equivalent to a change of 350 ml. The nature of osmoreceptors is the subject of much discussion. There may be two receptors; the classical osmoreceptor in an area without a blood brain barrier and a second receptor sensitive to the concentration of Na ions within cerebrospinal fluid. The two mechanisms by which detection of changes in osmolality may occur are:

1   Sensing a change in cell volume due to water depletion.
2   Sensing the ECF solute load.

In addition to this mechanism acting in close association with ADH there are other nonosmotic factors which are important in regulation of water balance (Table 1.4). If the effective ECF volume is severely depleted for example in intestinal obstruction this will override osmolality and produce fluid retention. If the cardiac output falls or intravascular volume depletion results in hypotension, this is detected by intrathoracic, carotid sinus and aortic baroreceptors, and mechanisms to increase blood pressure including fluid retention are stimulated. The renin angiotensin system is also activated and will result in sodium retention. These factors are discussed more fully in later chapters. In febrile states, water requirements are increased. Hypoxia impairs tissue perfusion and oxygenation which may result in impaired cell membrane activity (the sick cell syndrome) and fluid retention.

Table 1.4 Factors affecting thirst and ADH secretion.

Osmotic
Water depletion
Osmolar load

Nonosmotic
Volume receptors:
   arterial — carotid, aortic
   venous — right and left atrial, pulmonary venous
   renal, hepatic, central nervous system
Temperature
Hypoxia

The normal diurnal rhythm of ADH secretion

Antidiuretic hormone (vasopressin)

This is the most important hormone concerned with water balance. It is a nonapeptide with molecular weight 1084 and is synthesized in the supraoptic and paraventricular nuclei of the hypothalamus. It travels in vesicles with carrier protein (neurophysin I) to the posterior pituitary when it is discharged into capillaries. ADH acts on the distal tubule and collecting duct as has already been discussed, permitting increased reabsorption of water. Regulation of ADH secretion results in a range of urine osmolality (30–1200 mosmol kg^{-1}). A change in plasma osmolality of 1 mosmol kg^{-1} alters urine osmolality by 95 mosmol kg^{-1} by the ADH mechanism.

The normal pituitary content of ADH is 30 micrograms and its plasma half life is 16–20 minutes. At a plasma osmolality below 280 mosmol kg^{-1} ADH levels are undetectable. Within the normal range of plasma osmolality the full range of urine osmolality occurs with small changes in plasma ADH (0.4–5 pg ml^{-1}). The highest normal plasma osmolality (295 mosmol kg^{-1}) maximally stimulates ADH secretion, coincides with maximum urine osmolality and also stimulates thirst. As plasma osmolality rises from 280–295 mosmol kg^{-1} urine volume falls 10–20 times.

The osmoreceptor response to solutes is selective, that is to say, it is effective osmolality which is being regulated.

1 Glucose, urea and ethanol which readily cross the blood brain barrier do *not* stimulate ADH release although they increase plasma osmolality. In clinical hyperglycaemic states, therefore, the poor ADH response may aggravate polyuria and dehydration.

2 Sodium and mannitol which penetrate the blood brain barrier slowly are potent stimuli of ADH release. Therefore it is the

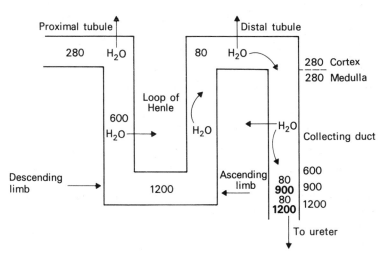

Fig. 1.12 The effect of antidiuretic hormone on urine osmolality. Numbers show osmolality (mosmol kg^{-1}); bold type = with ADH; medium type = without ADH.

gradient of osmolality between cells and ECF that stimulates ADH secretion or the reduction in cell volume which occurs as water moves out of the cell to restore isotonicity.

Alterations in osmolality influence base-line vasopressin but have little effect on the increase which occurs after haemorrhage. Vasopressin may influence renin activity normally but not its increase after haemorrhage.

Vasopressin release is in part controlled by endogenous opioid peptides especially beta-endorphin. In some, but not all, of the pathways leading to vasopressin release, endogenous opioids exert inhibitory control which may be by hypophyseal beta-endorphin.

In summary, therefore, regulation of ADH is mainly by osmotic stimuli but nonosmotic factors can affect the threshold for release and sensitivity of the response.

Other hormones affecting renal water secretion

1 Renal medullary prostaglandins.
Collecting duct tubular epithelium and interstitial cells can make prostaglandins which have three potent effects;
(a) Reduction in vasopressin dependent osmotic water permeability of the collecting tubular epithelium. The effects of vasopressin on isolated collecting ducts is mediated by cAMP and can be inhibited by prostaglandin E_1 which inhibits adenyl cyclase and inhibits postaglandin synthesis and is associated with enhanced vasopressin induced urinary concentration. This is probably not very important under normal circumstances but more so in shock and trauma.
(b) Enhanced medullary blood flow.
(c) Inhibition of Na and Cl absorption from the ascending limb of the loop of Henle.
Thus increased medullary prostaglandin content reduces medullary solute content and increases water excretion, ultimately antagonizing the effect of vasopressin. These actions of prostaglandins provide an integrated mechanism for the 'fine-tuning' of water excretion.
2 Catecholamines. Beta effects increase ADH release by a baroreceptor effect. Alpha effects reduce ADH release.
3 Thyroxine increases diluting capacity because of an increase in distal tubular sodium reabsorption or a reduction in ADH release.

4 Glucocorticoids increase ADH release.

5 Mineralocorticoids reduce diluting capacity. Nonosmotic stimuli may still cause ADH release.

6 Angiotensin has a variable response.

It is extremely difficult to separate water balance from sodium balance although an attempt has been made to do so. It is also difficult to discuss electrolyte balance in isolation from acid base regulation. Changes in total body water are reflected in altered plasma sodium concentration. Throughout this book changes will as far as possible be considered under the primary abnormality although inevitably some repetition will be necessary.

Chapter 2
Normal Sodium and Chloride Balance

Ye are the salt of the earth

Saint Matthew (Chapter 13)

Total body sodium (Na) varies between the sexes.

52–60 mmol kg^{-1} (adult male).

48–55 mmol kg^{-1} (adult female).

Therefore a 70 kg man contains 3600–4200 mmol Na. Since ECF (interstitial fluid plus plasma) forms about 20% of total body weight and extracellular Na concentration is 133–145 mmol l^{-1} then about 2000 mmol (50% of the total body Na) exists in the extracellular compartment. About 40% of the total body Na is found in the skeleton where 75% of it is very slowly, if ever, exchanged with that in other body fluids as it is absorbed on to hydroxyapatite crystals in dense bone (non-exchangeable Na pool).

Sodium therefore is the major extracellular cation. Intracellular Na concentration is low and varies between cell types:

3–4 mmol l^{-1} in muscle cells.

20 mmol l^{-1} in red blood cells.

EXCHANGEABLE SODIUM (Na_e)

This represents 65–70% of total body Na:

In females it comprises 41.7 mmol kg^{-1} body weight.

In males it comprises 40.1 mmol kg^{-1} body weight.

Exchangeable Na is made up of:

1 Extracellular Na.

2 Intracellular Na.

3 40–50% of total bone sodium.

It is this value that is measured by isotope dilution. ^{24}Na or ^{22}Na may be used for this purpose. There is considerable individual variation in Na_e and again values are higher in males than in females. Exchangeable Na from bone may partly replace Na lost from the ECF. As Na is the major ECF cation it contributes 86% of ECF osmolality (normally 280–295 mosmol kg^{-1}). Potassium plays a similar role in the ICF which is a much larger volume than the ECF. We have already seen that changing the osmolality in one body compartment is rapidly dealt with by the passage of water across cell membranes to produce isotonicity.

Administration of hypertonic saline

If hypertonic saline is infused into the circulation, the following effects are produced:
1 Plasma and interstitial fluid osmolality increase.
2 Water then leaves the ICF and enters ECF.
3 Isotonicity is restored.
The administered saline has stayed in the ECF but water has been added to maintain osmolality and the ECF has therefore been rapidly expanded at the expense of the ICF.

Administration of a large water load

If a large water load is administered into the circulation, it produces the following effects:
1 Plasma and interstitial osmolality are reduced.
2 Water leaves the ECF and enters cells.
3 Isotonicity is restored.
ICF volume is much larger than ECF volume, so that only a small increase in ECF volume will occur relative to the volume of water administered.

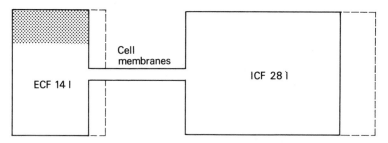

Fig. 2.1 The result of administration of a large water load. The dotted lines show the effect of a 10 litre water load. As water distributes freely between compartments osmolality is equally reduced in ECF and ICF. The shaded area denotes the intravascular compartment.

Loss of water without loss of solute

This is a rare situation and leads to reduction in the volume of ECF and ICF.

Loss of solute without water

Loss of solute from the ECF leads to a fall in osmolality with consequent passage of water from ECF to ICF, which may lead to swelling of cells. This situation will occur as administered glucose is metabolized to carbon dioxide and water, in effect water has been infused into the intravascular volume.

MEASUREMENT OF PLASMA SODIUM

There are two main methods: flame photometry and ion selective electrodes.

Flame photometry

Flame emission photometry

If a solution containing Na is squirted into a flame, a small number of atoms absorb energy so that one of the electrons moves further from the nucleus to a position of higher energy. As these atoms return to their original state they emit energy as light of specific wavelength. The amount of light emitted is proportional to the concentration of Na. Most machines compare this with an internal reference element (usually lithium).

Atomic absorption photometry

This is a more sensitive technique for research purposes.

Ion selective electrodes

Sodium glass electrode

When this electrode is in contact with a plasma sample a potential difference is set up proportional to the concentration of Na in the plasma. Very small volumes of plasma are required and the same system may be used for urine Na estimation. It is important that blood for plasma Na estimation is taken from a fasting, recumbent patient and from a limb with no intravenous infusion running, ideally without recourse to a tourniquet.

Plasma Na reflects the ratio between total ICF Na and extracellular water. An abnormal plasma Na therefore may be due to changes in either or both of these. There is a relationship between plasma Na and exchangeable Na and potassium.

$$\text{Na plasma } \alpha \; \frac{\text{Na}_e + \text{K}_e}{\text{TBW}}$$

NORMAL PLASMA SODIUM

The range for this value is 133–145 mmol l^{-1}.

Interpretation of this measured result

Hyponatraemia may be present in salt depletion or overload. This matter is discussed more fully in Chapter 7. There is no correlation between plasma and total body Na. Falsely low Na levels may occur if blood is taken near a vein with a hypotonic infusion running or if plasma contains excess lipid or protein.

The main difference between interstitial fluid and plasma is the protein content of the latter. Protein constitutes 6% of plasma volume but only 1% of interstitial fluid volume. Since Na is dissolved in the aqueous phase only, it might be expected that plasma Na would be less than interstitial fluid (ISF) Na whereas when measured in practice they are the same.

GIBBS DONNAN EFFECT

1 Protein is a non-diffusible anion present in plasma but insignificantly in interstitial fluid.
2 Since electrical neutrality must be maintained the diffusible ions must distribute *unequally* across the capillary membrane so that the total number of cations equals the total number of anions on each side.

3 Within the capillary containing non-diffusible protein anion, the concentration of diffusible anions must be less than outside the capillary (ISF) and the concentration of cations must be more.

4 Effectively, therefore, diffusible anions such as chloride are in lower concentration in plasma than ISF and diffusible cations such as Na are in higher concentration (6–7 mmol l^{-1} greater) in plasma than in ICF.

However, Na concentration is measured in mmol l^{-1} of plasma *not* per litre plasma water and hence achieves a value almost identical to that in ISF (see Fig. 1.2).

Pseudohyponatraemia

Sodium is present only in the aqueous phase of plasma so that if plasma contains an abnormal amount of lipid contracting the water volume then the measured sodium will be spuriously low.

Such a situation exists in nephrotic syndrome and diabetes mellitus. If in 1 litre of plasma there are 800 ml water and 200 ml lipid with a measured Na of 120 mmol l^{-1}, then the calculated true plasma sodium is 150 mmol l^{-1}. This phenomenon may have dangerous results if not recognized and if an inappropriate sodium load is given to correct the hyponatraemia. Measurement of urine sodium may help to diagnose this situation. In true hyponatraemia urine Na and Cl will be low. This topic is further discussed in Chapter 7. Use of an ion selective electrode which measures activity of ions in plasma water overcomes this problem to a large extent being relatively unaffected by increases in lipids or proteins which reduce the plasma water space. Such machines are of considerable help provided they do *not* dilute the plasma before presenting it to the electrode.

SODIUM BALANCE

Intake

1 Food.   **2** Drinks.   **3** Added salt with food.

Obviously the intake is very variable depending on diet and taste. Most of the daily Na intake is due to added salt with meals, hence restricted Na diets are not difficult to plan. The highlanders of

New Guinea consume less than 10 mmol per day and remain healthy, as do a small number of people consuming up to 200 mmol of Na daily. Optimum intake, however, is about 75 mmol per day. 70–100 mmol daily of Na is suggested as a baseline replacement. If Na intake is very low minor disturbances of gastrointestinal function may lead to hyponatraemia and dehydration. This is akin to being at the edge of a precipice. The true minimum Na intake should probably be 20 mmol daily.

The sodium content of certain drugs, especially antibiotics, can be very high and represent an important increase in total sodium intake in patients with heart failure (amoxycillin 2.6 mmol-Na g^{-1}, ticarcillin 4.7 mmol Na g^{-1}).

Output

1 Faeces and skin. 5–10 mmol Na are lost daily by these routes.
2 Sweat. Under normal circumstances the losses are small (less than 25 mmol daily). In the presence of fever and visible sweating due to muscular activity, obesity or high environmental temperature, losses may be much higher.
3 Urine. The kidney normally maintains Na balance by excreting all the Na which is in excess of requirements. The normal kidney can probably excrete over 300 mmol Na daily except in Na retention states although sodium excretion is reduced in the elderly. In conditions of salt loss or deprivation, however, it is very efficient in conserving Na and then daily urinary losses may be only 1 $mmol^{-1}$ after a few days. On an average Western diet urine K:Na ratio is 1:2.

The renal regulation of salt balance is not an immediate phenomenon so that small daily differences may occur between intake and output. Large fluctuations in salt intake may require 3–5 days for the kidney to establish a new steady state of Na output.

MEASUREMENT OF URINE SODIUM

This is becoming increasingly common in ill patients. However, the value of such measurements must be interpreted with caution. An isolated urine Na measurement is of very limited value as it is dependent on dietary Na and recently administered diuretics, the normal range of which is very wide. This is discussed more fully in Chapter 7. A 24 hour urine Na may demonstrate the renal ability to excrete Na and a low urine Na in oliguric states indicates a low perfusion state.

NORMAL REGULATION OF SODIUM BALANCE

Salt hunger does develop in states of need, for example adrenal insufficiency, very low salt diets and recognition of the threshold for salt may be increased in hypertension.

We have seen that Na intake is very variable depending on diet and taste, how then is excretion regulated to compensate for variable losses via the gastrointestinal tract and sweat? This is effected by the kidney. There is some evidence that a gut hormone may be involved. Administration of hypertonic saline orally is associated with a much higher rate of excretion of Na than administration of an equivalent dose intravenously.

RENAL REGULATION OF SODIUM BALANCE

The glomeruli within the kidney filter plasma, allowing water and electrolytes but not protein, through into the tubule. The working of the countercurrent multiplier has been considered in Chapter 1. Glomerular filtrate resembles plasma initially but during its course through the tubule large amounts of water and variable amounts of electrolyte are reabsorbed so that by selective active and passive control mechanisms the renal tubule controls electrolyte balance within the body. In addition to reabsorption, tubular secretion is important for electrolytes such as potassium.

There is still no convincing demonstration of any Na receptor in the body nor much evidence of how the Na status of the body is relayed to the kidney. How then does the kidney detect changes in body Na content? The mechanism is likely to include changes in ECF volume in the following way.

When ECF volume falls the distending pressure in the large arteries and carotid sinus falls; arterial blood pressure may also fall. As ECF volume falls the sympathetic nervous system is reflexly stimulated, which produces the following:

1 Catecholamine induced increase in cardiac output.
2 Peripheral constriction, especially in skin, splanchnic and renal vessels.
3 Stimulation of the renin angiotensin system.

The kidney therefore may be affected by changes in its blood pressure or blood supply mediated by the mechanisms listed above. Other volume receptors in addition to the carotid sinus are found in the left atrium, the right atrium and juxtamedullary apparatus. Volume expansion stimulates both the high and low pressure baroreceptors to increase Na excretion by a mechanism independent of the constitution of the blood. There are three possibilities for relay of information from the baroreceptors to the kidney:

1 Circulating natriuretic hormone.
2 Dopamine.
3 Decreased renal nerve activity.

CONTROL OF SODIUM EXCRETION

1 Glomerular filtration rate (GFR).
2 Aldosterone.
3 Natriuretic factors.

GFR

Reduction in GFR reduces urinary Na excretion and vice versa. This mechanism can, however, be overridden. In animals a large salt load given in the presence of an experimental reduction in GFR will still produce a marked Na diuresis (natriuresis). Renal nervous activity is one factor controlling GFR.

Aldosterone

This mineralocorticoid hormone is secreted by the zona glomerulosa of the adrenal cortex. It stimulates renal tubular reabsorption of Na and decreases the Na content of sweat. Aldosterone release is stimulated by angiotensin (Fig. 2.2).

Stimuli such as a reduction in renal perfusion pressure increase renin secretion from the juxtaglomerular cells surrounding the

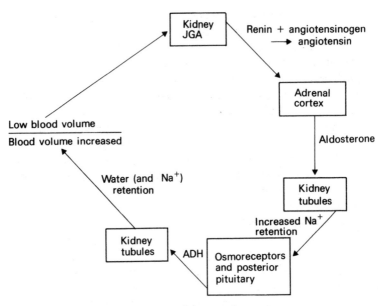

Fig. 2.2 Effects of aldosterone on sodium retention.

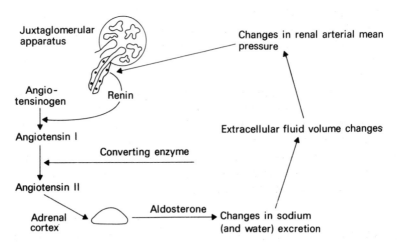

Fig. 2.3 The renin angiotensin system.

renal afferent arterioles as they enter glomeruli. Renin converts angiotensinogen (a circulating alpha-globulin) into angiotensin I. Converting enzyme from the lung then converts this into angiotensin II, an octapeptide which in turn stimulates aldosterone release (Fig. 2.3). Angiotensinogen concentration normally remains constant. This is not the rate limiting step in the renin cascade but in patients taking oral contraceptive drugs angiotensinogen increases three-to-fivefold. This may increase the rate of generation of angiotensin II within the body. Other peptide fragments (angiotensin III) may also be important in control of aldosterone secretion.

The role of this aldosterone secreting mechanism is probably mainly in defence of intravascular volume rather than primarily regulation of Na excretion. In salt deprived animals chronic infusion of converting enzyme inhibitor causes decreased arterial blood pressure, reduced glomerular filtration rate and aldosterone concentration but increased urinary sodium excretion. Infusion of angiotensin II *in addition* causes increase in the blood pressure and renal function to control levels. The renin angiotensin system therefore plays a major role in regulation of blood pressure, renal haemodynamics and sodium excretion during sodium deprivation probably by a direct effect on angiotensin I rather than changes in plasma aldosterone.

The site of action of aldosterone is probably in the cortical and medullary collecting ducts. It is relatively slow to act and concerned with fine regulation. Aldosterone secretion increases on rising from the supine to the upright position and Na secretion is inhibited. However, this mechanism develops too rapidly for the increase in aldosterone to be totally responsible. Aldosterone secretion takes 10–30 minutes to exert any effect on Na excretion even if injected directly into the renal artery. This may represent the time required for increased enzyme secretion. Neither can this mechanism account for the regulation of Na excretion completely since if a large salt intake is given with a high dose of exogenous mineralocorticoid an *escape* phenomenon occurs and salt excretion eventually increases so that all the salt intake in excess of normal losses is in fact excreted. Before this mineralocorticoid escape there may be some ECF expansion prior to the new steady state but no oedema. This escape may occur because of a third factor (natriuretic factor) which will be discussed below.

Renal prostaglandins (PG), found mainly in the medulla, are undoubtedly important in the regulation of renin release. Stimulation of renal PG synthesis by frusemide increases renin release, whereas PG synthesis inhibitors such as indomethacin reduce such renin release. In certain disorders with increased renin production

such as Bartter's syndrome the increased renin is thought to be
secondary to increased renal PG synthesis. Large doses of i.v.
angiotensin cause natriuresis in which the initial effect is release of
PGE, then kallikrein which activates release of natriuretic hormones (Fig. 2.4). Large doses of PGE may exert their effect by
increasing blood flow but the role of PG in sodium excretion
remains to be completely elucidated.

TRADITIONAL VIEW OF SODIUM AND CHLORIDE EXCRETION

Sodium is filtered in the glomerulus, a large amount appearing in
filtrate. It is, however, transported out of the tubule either actively

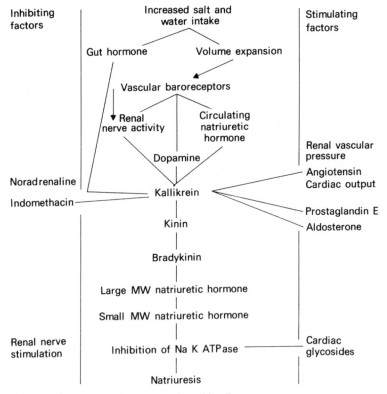

Fig. 2.4 Response to volume expansion with saline.

or passively following active chloride transport (see Chapter 1). The true distal tubule is capable of Na reabsorption independent of aldosterone whereas Na transport in the cortical collecting system is aldosterone dependent. Very large amounts of Na are filtered daily (26 000 mmol). If tubular reabsorption stays constant, a small variation in glomerular filtration rate produces a marked change in Na excretion. In fact, tubular reabsorption of Na varies with GFR but never completely compensates for changes in filtered load, hence the importance of GFR in affecting the amount of Na excreted. In the tubule most of the Na is reabsorbed with chloride but some is exchanged for hydrogen and potassium by tubular secretion (see Fig. 3.4).

Figure 2.4 illustrates the response to volume expansion with saline. There is a good correlation between the rate of Na excretion and urinary dopamine excretion. This dopamine is synthesized within the kidney and acts on specific dopamine receptors

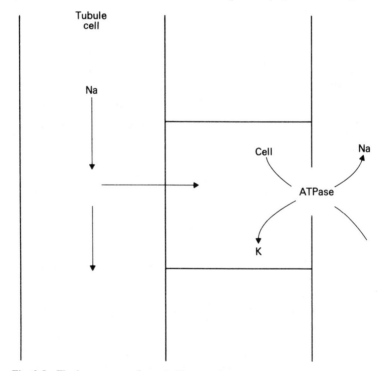

Fig. 2.5 Final common pathway in Na excretion.

activating the kallikrein/kinin system. This system plays a key role in control of Na excretion and it is at this level that blood pressure and angiotensin exert their effects. Positive pressure to the lower half of the body raises right atrial pressure, increases sodium excretion and urine flow. If the action of dopamine is blocked with domperidone the increase in Na excretion does not occur.

Other factors act via prostaglandin E to increase kallikrein release. Noradrenaline has two effects: inhibition of the prostaglandin stimulation of kallikrein and a direct action on renal tubular cells to stimulate Na/K ATPase. Stimulation of the kallikrein kinin system by dopamine results in production of natriuretic hormone which inhibits renal Na/K ATPase to produce natriuresis (Fig. 2.5). Na/K ATPase is the common pathway then in Na excretion. Stimulation of this enzyme is the major Na retaining mechanism and inhibition results in Na excretion or natriuresis.

Natriuretic factors

In the early 1980s, Gruber and colleagues found circulating digitalis-like factors in plasma loaded dogs who were not receiving digoxin. Similar substances have been found in normal people and levels may rise with increased salt intake and age. These factors inhibit Na-K ATPase and promote natriuresis. Leucocytes in hypertensive patients have depressed sodium pump activity leading to increased intracellular sodium concentration which may thereby increase intracellular calcium concentration and vascular tone.

In patients with essential hypertension the blood pressure is sensitive to sodium loading and there is reduced responsiveness to the renin angiotensin system. Concentration of natriuretic factors varies inversely with plasma renin activity and aldosterone as might be expected of a natriuretic hormone stimulated by increased body sodium, ECF volume or both. Atrial peptides therefore inhibit renal renin secretion and adrenal aldosterone secretion. The sights of receptors detecting plasma volume changes and natriuretic hormone synthesis are now being elucidated. Specific granules have been isolated in atrial muscle from an area which, if distended, releases material having diuretic action.

Atrial natriuretic hormone is a peptide (ANP) which increases excretion of magnesium, calcium and phosphate in addition to sodium and water in healthy volunteers. Administration is accompanied by a fall in blood pressure due to inhibition of the response to a range of vasoconstrictors including noradrenaline and angiotensin II. ANP is raised in children with a variety of cardiopulmonary diseases probably due to the increased preload resulting in atrial distension. It is also increased in almost all patients with heart failure.

However, since i.v. injection of ANP in healthy humans results in immediate, striking diuresis and natriuresis, this increased ANP must be a compensatory response to volume expansion. Plasma levels of 3'5' guanosine monophosphate (cyclic GMP) are high in congestive cardiac failure. ANP increases accumulation of cGMP in vascular smooth muscle of animals, therefore the higher plasma levels of cGMP in congestive cardiac failure may be explained by the increased ANP.

Plasma ANP concentration is increased in essential hypertension, again as a compensatory response unrelated to age. The head down posture similarly increases ANP, an effect related to the increase in central venous pressure thus produced. Patients with cirrhosis have reduced natriuretic and diuretic response to volume expansion but increased basal ANP with a blunted response to its action. The hypotensive effect of ANP further reduces hepatic blood flow. The major effect of ANP seems to be inhibition of sodium reabsorption from the distal tubule especially the collecting duct, although the vasodilator action may be important by increasing glomerular filtration fraction. The right atrium seems to have 3–4 times the ANP activity of the left. It is clear that much remains to be explained about all these interactions.

OTHER FACTORS CONCERNED WITH SODIUM EXCRETION

Colloid osmotic pressure in peritubular capillaries surrounding the proximal tubule controls reabsorption of Na and water in this region. Proximal tubular Na reabsorption occurs in three stages.

1 Sodium ions diffuse passively from the tubular lumen into proximal tubular cells.

2 Sodium or chloride ions are pumped out of the cells into intercellular spaces between the cells producing local hypertonicity. Chloride or sodium and water follow passively to render isotonicity.

3 Isotonic fluid passes into peritubular capillaries and leaks back into the tubular lumen via tight junctions between tubular epithelial cells.

Na and water reabsorption in the proximal tubule is equal to the total volume of isotonic resorbate formed within the intercellular spaces minus the volume of that which diffuses back into the tubular lumen.

This suggests that the net proximal Na and water reabsorption is increased if peritubular capillary blood has a high colloid osmotic pressure. Infusion of crystalloid solutions will lower capillary colloid osmotic pressure and result in reduced reabsorption.

In congestive cardiac failure excess salt and water retention may result in oedema. This may be explained by an increase in peritubular plasma protein and hence colloid osmotic pressure again resulting in increased Na and water reabsorption.

Hydrostatic pressure

As in capillaries elsewhere the hydrostatic pressure difference across the peritubular capillary wall is an important factor in reabsorption. A high capillary hydrostatic pressure will reduce reabsorption.

Blood flow

If the peritubular blood flow rate is low this may allow earlier dilution of the peritubular capillary blood by resorbate, an earlier reduction of colloid osmotic pressure and reduced net salt and water reabsorption.

Renal nerves

Denervation of the kidney is associated with a large increase in urinary sodium excretion. Mechanisms include a fall in arteriolar resistance, increased filtration fraction, directly and via control of renin, changes in renal blood flow and tubular reabsorption.

The physiological role of the adrenergic nervous system in regulation of salt and water is not completely understood but it may be important in situations requiring acute conservation of salt whilst being only a minor factor in normal control of sodium excretion. Noradrenaline infusion produces marked oliguria which can be antagonized by an alpha-adrenergic blocking drug.

Insulin

Increases in plasma insulin within the physiological range stimulate sodium reabsorption by the distal tubule and this is independent of other hormones. This effect may be important in the sodium wasting of poorly controlled diabetes and sodium retention during refeeding.

In the newborn, the increased umbilical adrenaline after vaginal delivery compared to Caesarian section stimulates Na and water reabsorption from the lung by a beta-2-adrenergic effect.

SODIUM REABSORPTION ALONG THE NEPHRON

Proximal tubule

1   Normally 60–70% of filtered Na is reabsorbed in this region.
2   All proximal reabsorption is isotonic.
3   The composition of tubular fluid changes along the tubule. Active reabsorption of bicarbonate occurs early in the proximal tubule in excess of water reabsorption. This results in a reduction in bicarbonate concentration in tubular fluid and in exchange, chloride concentration is increased.
4   Na reabsorption in this area may be active or passive (Na ions moving down a favourable electrical gradient generated by active or passive chloride diffusion). Much of the glucose and amino acid reabsorption is Na coupled.

Ascending limb of the loop of Henle

20–30% of filtered Na is reabsorbed here. The traditional view states that Na is actively pumped out of this segment to produce the hypertonic interstitium required for operation of the countercurrent mechanism. More recent concepts suggest that it is chloride which is actively pumped out, facilitated by Na-K ATPase and Na then follows passively along a favourable concentration gradient.

Distal tubule

1 The early part of this segment is the true distal tubule.

2 The latter part is the beginning of the collecting duct in the cortex. In the true distal tubule active Na reabsorption takes place independently of aldosterone whereas Na reabsorption in the cortical collecting duct is aldosterone dependent.

3 5–10% of the filtered Na is reabsorbed in the distal tubule either in association with the chloride or under the influence of aldosterone in exchange for potassium or hydrogen ions.

Collecting duct

Na reabsorption here is extremely variable depending on the action of aldosterone and antidiuretic hormone and on the patient's Na status. This area is extremely important in regulating Na balance and urinary Na output. In salt depletion marked Na reabsorption occurs but in salt overload no Na is reabsorbed and recent suggestions point to Na secretion in the collecting duct.

Therefore urinary excretion of sodium and water varies in response to dietary intake and effective blood volume. Volume receptors sense the adequacy of filling of the circulation and urinary sodium excretion is modified by changes in both glomerular filtration rate and tubular reabsorption. Sodium excretion will also be influenced by the peritubular conditions, renal nerves and a variety of hormones. The whole mechanism maintains ECF almost constant. It now seems that signals from the oropharynx contribute to a diuresis occurring after drinking isotonic fluids and that diuretic response occurs with similar latency to that of a water diuresis and maximal at about 50 minutes from the start of drinking.

CHLORIDE BALANCE

This is very similar to Na balance. Bromide can interfere with chloride estimation and with certain instruments give a falsely high recording. Bromide is still contained in some sedative drugs and tends to accumulate in the blood.

Chloride intake

70–210 mmol of chloride are ingested as sodium chloride or potassium chloride in food and drink daily via the gastrointestinal tract. The minimum daily requirement of chloride is 75 mmol. Chloride is absorbed in the intestine with Na and potassium by an unknown active process.

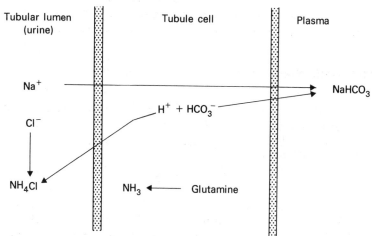

Fig. 2.6 Bicarbonate reabsorption in the tubule.

Output

Chloride is lost:
1 In the sweat with Na in a concentration of about 15 mmol l^{-1}.
2 In gastric juice 90–150 mmol l^{-1}.
3 In bile, pancreatic and intestinal fluids 50–100 mmol l^{-1}.
In the renal tubule chloride excretion varies with the needs of the body for bicarbonate. Chloride is excreted with ammonium ions to eliminate hydrogen ion in exchange for Na (Fig. 2.6).

This occurs mainly in the proximal tubule. In the ascending limb of the loop of Henle, chloride is reabsorbed with Na (Fig. 1.9). Regulation of chloride is passively related to Na but inversely related to plasma bicarbonate. Aldosterone therefore indirectly influences chloride levels.

Chapter 3
Normal Potassium Balance

*Potassium is of the soil and not the sea,
of the cell and not the sap.*

W. O. FENN (1940)

Most of the potassium (K) within the human body is within the cells.

A 70 kg man contains 2900–3500 mmol of K (about 50 mmol kg^{-1} body weight).

Normal plasma K is 3.5–4.8 mmol l^{-1}.

Intracellular K is about 150 mmol l^{-1}.

Therefore only about 60 mmol of K (2% of the total body K) is found in ECF. In effect therefore large changes may occur in total body K without a significant effect on plasma K concentration. Plasma K is less well buffered than any other major ion.

EXCHANGEABLE K (K_e)

This is the pool of K within the body which comes into equilibrium with the radioactive isotope ^{42}K within 24 hours. In humans virtually all K is exchangeable with two exceptions:

1 K in erythrocytes exchanges only slowly.

2 Some of the K in bone equilibrates only very slowly with ^{42}K.

Exchangeable K in males is 42–48 mmol kg^{-1} body weight.

Exchangeable K in females is 34–38 kg^{-1} body weight.

Thus there is a significant difference between the sexes: the highest value of total body K being in muscular males. In both sexes total body K declines with age.

K_e has been related to Na_e and to total body water. In health the relationship Na_e/K_e is approximately 0.85 in males and 1.0 in females. This relationship rises in illnesses such as sepsis, heart failure and trauma where cell membrane function may be impaired, to 1.5.

The following relationship has been demonstrated between Na_e, K_e and total body water.

$$Na_e + K_e(mmol) = 163.19 \, TBW(l)-69.$$
$$r = 0.99 \, (P < 0.001).$$

Membrane polarization

Despite the fact that a very small proportion of total body K is present in ECF it is regulation of plasma concentration which is very important for maintaining membrane polarization.

Cells within the human body exist in a state of polarity. There is a resting potential difference of 50–90 mV between the inside and outside of the cell (the inside being negative to the outside). The magnitude of this potential is proportional to the ratio of the concentration of the most permeable ions on each side of the membrane. In the resting state K is the most permeable ion and therefore the difference in its concentration between the inside and the outside of the cell is the main determinant of the membrane potential. Na is much less permeable in the resting state than K and under normal circumstances does not leak into the cell. In excitable tissue depolarization above a threshold of about -60 mV results in an action potential which results in passage of Na into the cell, the membrane potential then falls and depolarization occurs. In cardiac muscle a second slow inward current is responsible for calcium entry into the sarcolemma initiating contraction. The magnitude of an action potential depends on the distribution of Na across the cell membrane. Following this event the sodium pump restores Na to its extracellular position and K equilibrium is restored. Outward K movement is activated slowly on depolarization and deactivates rapidly after repolarization has taken place.

MEASUREMENT OF POTASSIUM

Exchangeable potassium

This is measured using the radioactive isotope ^{42}K which equilibrates with 90% of total body K within 24 hours. ^{42}K is given intravenously or orally. When blood is collected for measurement of ^{42}K activity, plasma should be separated from red cells within 2 hours of sampling.

Total body potassium

This requires a whole body counter to measure the emission of naturally occurring ^{40}K and is therefore of limited clinical application. However, it has been recently demonstrated that there is a good correlation between total circulating blood cell K content and whole body K. Potassium concentration can be measured directly in erythrocytes, leucocytes and skeletal muscle cells.

Plasma potassium

This may be measured by:
1   Flame photometry.
2   Ion specific electrodes.
These methods have been discussed in Chapter 2 under measurement of plasma Na.

Ion selective electrodes are being introduced into high dependency areas such as cardiac surgical intensive care units where they provide rapid accurate estimation of K.

Interpretation and value of such measurements

A measurement of total body K as a prediction of K depletion or excess is very limited since the range of normal is wide and a 20% change is required before we can be assured of abnormal total body K. Serial changes in one individual are much more informative. These same restrictions also apply to exchangeable K. Many forms of illness lead to a catabolic state in which endogenous protein from muscle is broken down for energy releasing a large amount of K into plasma which is then excreted in the urine. Artefacts can occur during blood sampling. The same precautions should be taken as suggested for Na measurement. It is most important that haemolysis is avoided otherwise a spuriously high K may result by leakage from red blood cells. Samples should not therefore be shaken and plasma should always be separated within 2 hours of sampling. Refrigeration worsens haemolysis. In the course of blood clotting K is released from cellular components of blood but in the normal patient this is without clinical significance. Some blood disorders in which leucocytes or platelets are abnormally fragile may result in a spuriously high K value. Even severe muscle exercise may raise plasma K. As less than 3% of total body K is found in the ECF the relation between plasma K and total body K is tenuous. Total body K depletion can occur with a normal or raised plasma K. Although a low plasma K does suggest depletion this must be severe before the plasma level reflects the change in total body K. Over 10% of total body K (400 mmol) must be lost before there is any reflection in plasma K.

Factors which have an effect on membrane transport have a marked effect on plasma K whilst total body K remains unchanged.

These factors are:

1 Acid base state.
2 Diuretics.
3 Carbohydrate metabolism.
4 Hypoxia.
5 Digitalis.

Their effects are mediated by a change in ICF to ECF ratio. Hypertonic infusions increase plasma K by 0.1–0.6 mmol l^{-1} for each 10 mosmol kg^{-1} increase in osmolality, perhaps by increasing the ICF/ECF K flux.

It is unwise to treat an isolated abnormality of plasma K without consideration of the full clinical picture. In an acute disturbance it is the plasma concentration which is crucial regardless of the total body status since plasma concentration has the greatest effect on the membrane potential. In any event if an acute K disturbance exists or rapid replacement is contemplated continuous electrocardiographic (ECG) monitoring is invaluable to show minute by minute changes in transmembrane K gradient. Although strictly this constitutes an abnormal situation the changes in the ECG will be illustrated here as they show the importance of the K level on membrane electrophysiology.

When plasma K is less than 3.0 mmol l^{-1} the T wave becomes broader and flatter and the U wave may fuse with the T wave (Fig. 3.1(b)). As plasma K rises the QRS widens and tall peaked T waves appear (Fig. 3.1(c)). The P wave then becomes less prominent and may disappear (Fig. 3.1(d)). The ECG then takes on the appearance of a sine wave. Complete heart block, continued widening of the QRS complexes and ultimately asystole may occur.

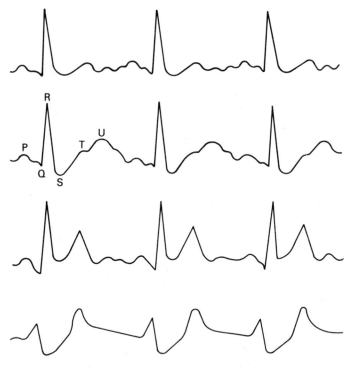

Fig. 3.1 Electrocardiographic changes at various levels of plasma potassium. (a) normal ECG; (b) hypokalaemia; (c) hyperkalaemia; (d) severe hyperkalaemia.

POTASSIUM BALANCE

Intake

The daily K intake is between 50–200 mmol varying widely with dietary food and fluid habits. It is almost impossible that a diet containing sufficient energy will be deficient in K but, as patients are prepared for surgery by starvation, K intake immediately falls and a situation of negative K balance occurs. Potassium intake normally greatly exceeds requirements and whereas most of the Na in the diet is in the form of added salt, K is present in all living cells and therefore fruit, vegetables and meat contain 100 times as much K as Na so that K free diets are totally impractical.

Output

Obligatory losses:
1 Skin and normal stools: 15–20 mmol daily.
2 Urine: 10–20 mmol daily (even on a K free diet).
Potassium cannot be so efficiently conserved by the kidneys as can Na although there is a large reserve for the excretion of K. Thus a minimum of 40 mmol K are required to cover basal daily losses.

Postoperative situation

After uneventful surgery 50–100 mmol K may be lost in the urine in the first 48 hours. If in addition the surgery is severe or associated with postoperative complications such as sepsis, wound breakdown or prolonged ileus this K loss continues throughout the period of catabolism in conjunction with muscle breakdown and negative nitrogen balance. Intake is likely to be reduced as K is often withheld in the immediate postoperative period. Therefore considerable deficits of K can occur within a few days of an operation if no K is administered and it must be remembered that this may not be reflected in a low plasma K. Evaluation of K balance may be helped by measurement of 24 hour urinary K losses. If the loss is less than 60% of the K intake then a total body K deficit can be presumed to exist. It should be remembered that very significant losses of K can occur from the gastrointestinal tract especially if nasogastric suction or prolonged diarrhoea are problems. This is discussed further in Chapter 8. Intravenous administration of K salts results in rapid equilibration. If the cardiac output is normal 95% equilibrium occurs in a single transit through the circulation.

Regulation of potassium balance under normal circumstances

In man mechanisms to prevent K overload are much more highly developed than those to prevent K loss and respond directly to changes in plasma K. In conditions of K loading or Na deprivation, the increased K secretion is due to increased capacity of the colonic cells to secrete K and associated increase in Na/K ATPase activity in both cases. Cells of the distal nephron also increase their rate of secretion and number of K pumps. This same effect occurs in other conditions such as chronic renal failure, hyperaldosteronism and glucocorticoid administration.

Since it determines membrane polarization it is the plasma K concentration that requires precise regulation by one of two mechanisms:

1   Changes in K distribution across the cell membrane.
2   Renal excretion. Except in Addisonian crisis life threatening hyperkalaemia rarely arises in the presence of normal renal function.

Renal excretion is responsible for long term regulation of K balance whereas changes in distribution across the cell membrane account for acute alterations in plasma K. Decreased K intake or excessive loss is 'buffered' by release of K from cells into ECF.

Mechanisms for acute redistribution of potassium

1   Autonomic nervous system.
2   Pancreatic hormones.
3   Acid base status.

Autonomic nervous system

Catecholamines have a biphasic effect on potassium balance and a physiological role in internal K balance.

1   Adrenaline increases plasma K by an alpha adrenergic effect probably stimulating hepatic K release.
2   Following this adrenaline produces a sustained decrease in plasma K by a beta adrenergic effect which is independent of insulin due to K uptake by muscle. This effect can be blocked by beta 2 or nonselective beta adrenergic blockade but not by beta 1 blockade. Transient hypokalaemia in recently hospitalized patients may reflect their anxiety. This reduction in K also occurs with beta 2 agonists such as fenoterol and salbutamol, an effect which may be particularly dangerous during the treatment of asthma, especially if K is already low due to steroids, diuretics or if they are used in combination with aminophylline. More often asthma deaths are due to undertreatment so this hypokalaemic effect must be kept in perspective. Stimulation of alpha adrenoreceptors impairs extrarenal disposal of an acute K load (in oppo-

sition to beta stimulation) and this alpha adrenergic effect may act to preserve plasma K or contribute to hyperkalaemia under certain circumstances such as vigorous exercise.

One of the ganglion blocking drugs currently used for inducing hypotension has been shown to reduce plasma K for the period of intraoperative hypotension.

Pancreatic hormones

INSULIN

An increase in plasma K which is well within the normal physiological range will stimulate insulin release resulting in uptake of glucose and K into cells from ECF (Fig. 3.2).

GLUCAGON

An increase in plasma K also stimulates release of glucagon which raises plasma glucose.

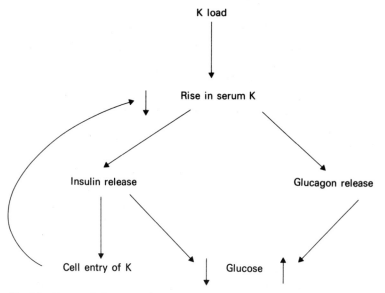

Fig. 3.2 Pancreatic hormones in acute redistribution of potassium.

The net effect of these two hormones is uptake of K into cells with minimal effect on blood glucose. This K uptake into cells with glucose, facilitated by insulin, is one of the most rapid means of reducing a high plasma K and is used therapeutically.

The effectiveness of these hormones has been demonstrated in pancreatectomized dogs given a K load. Survival time is dramatically reduced unless the K load is accompanied by insulin and glucagon. Conversely in diabetics insulin induced hypoglycaemia may be associated with dangerous hypokalaemia.

Acid base status

Intracellular pH has an important effect on K ion activity. An intracellular alkalosis (low hydrogen (H) ion concentration) promotes influx of K into the cell to restore electrical neutrality. Intracellular acidosis (high H ion concentration) conversely inhibits K influx into cells and favours movement of K out into ECF.

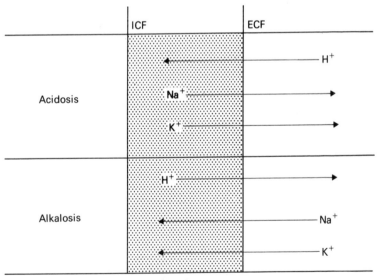

Fig. 3.3 Potassium exchange in acute disturbance.

The relationship between pH and K in plasma is as follows: for each rise or fall of 0.1 pH unit there is an average fall or rise in plasma K of 0.63 mmol l^{-1}.

Renal regulation of potassium balance

There is a good correlation between total body K content and K excretion by the kidney but the correlation between plasma K and urinary excretion is poor. The kidney has a large excretory reserve for K.

There are no regulatory mechanisms controlling K intake, absorption or proximal renal tubular absorption. 85–90% of filtered K is reabsorbed in the proximal renal tubule and further reabsorption takes place in the loop of Henle. Therefore the glomerular load of K is all reabsorbed by this time and K excretion is entirely due to distal tubular mechanisms. In the absence of renal failure the glomerular filtration rate (GFR) normally has little effect on K excretion. The only exceptions to this are in the following abnormal circumstances:

1 Osmotic diuresis.
2 Massive ECF volume expansion.
3 Administration of frusemide.

In these cases delivery of a larger than normal K load to the distal renal tubule contributes to urinary excretion of K.

Tubular fluid to plasma K concentration ratios increase from 0.2–5.0 along the distal tubule. The collecting ducts contribute little to K excretion. Net reabsorption of K is only seen after severe dietary restriction. There is evidence of an active reabsorption mechanism which is inhibited by ouabain. Net excretion can be increased by a diet rich in K.

Distal tubular regulatory mechanisms

The main factors influencing K excretion are:

1 Aldosterone.
2 Acid base status.
3 Na/K ATPase activity.
4 Flow rate through the distal tubule.

ALDOSTERONE

Aldosterone is the most potent of the naturally occurring mineralocorticoid hormones and is concerned in the regulation of Na and K balance. A K load increases aldosterone production and

K depletion diminishes it. This effect is probably quite independent of changes in the volumes of body fluids and the renin angiotensin system. The effect of aldosterone is to conserve Na and promote excretion of K (see also Chapter 2). An increase in plasma K of less than 0.5 mmol l^{-1} will increase aldosterone secretion whilst a pre-existing high K intake will magnify the response to an acute K load. Aldosterone therefore is a powerful controller of ECF K concentration such that a sevenfold rise in K load only increases plasma K by 2–2.5%.

Mechanism of action of aldosterone. The distal tubular lumen becomes progressively more electronegative as Na is actively reabsorbed. This favours passive movement of K into the lumen of the tubule to restore electrical neutrality. In addition to this passive movement of K into the tubular lumen K is also actively secreted into the tubular lumen (Fig. 3.4). This effect is maximal in the late distal tubule.

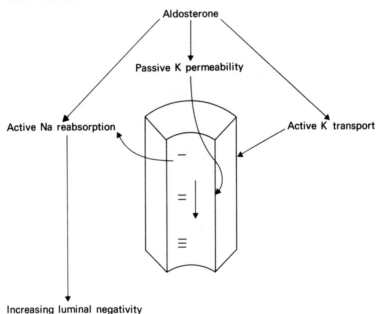

Fig. 3.4 Action of aldosterone on the distal renal tubule.

Aldosterone enhances both these effects:
1 It increases passive K permeability.
2 It increases the active K entry from peritubular fluid into the

distal tubular cells. If excess aldosterone activity is present K depletion and hypokalaemia may occur. Then Na retention occurs and Na ions partly replace the intracellular K deficit.

ACID BASE STATUS

In the presence of an alkalosis, K secretion is increased and in acidosis it is decreased. Within the distal tubule both H ions and K ions compete for excretion in exchange for Na reabsorption.

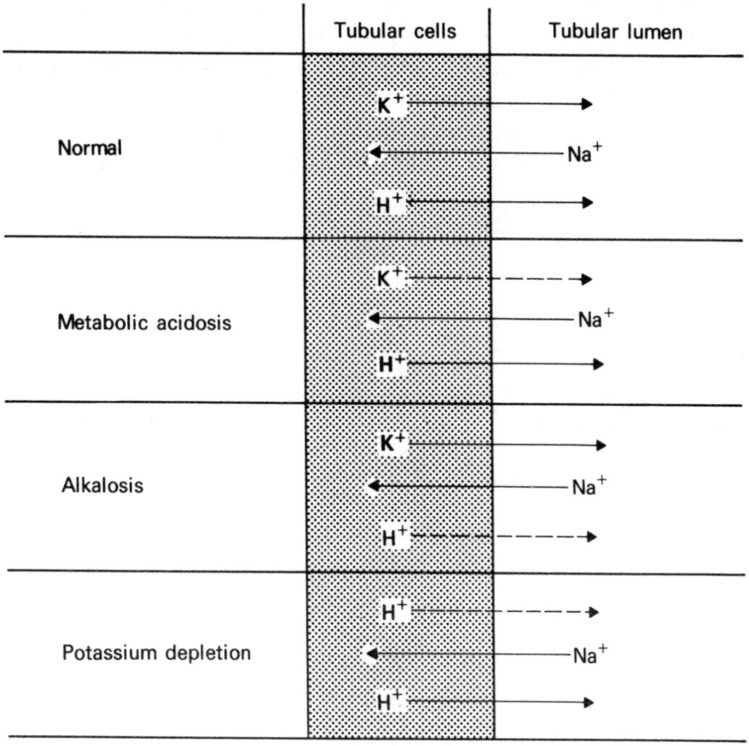

Fig. 3.5 Potassium flux in the distal tubule in acid base disturbance.

Metabolic alkalosis. When a metabolic alkalosis exists with a deficit of H ions, intracellular K increases in all body cells including those in the distal tubule. The rate of K excretion is enhanced resulting in total body K depletion. This depletion will result in a fall in intracellular K. Therefore the tubule exchanges Na for K ions in order to conserve H ions. At a later stage K secretion will diminish and more H ions will be excreted. This results in paradoxical aciduria in alkalosis with K depletion.

Metabolic acidosis. In metabolic acidosis there is an increase in H ion excretion and the rate of K excretion is markedly diminished. These changes are illustrated in Fig. 3.5. Intracellular K concentrations are directly related to the rate of distal tubular K excretion. Changes in peritubular K uptake can partly explain the effects of a change in acid base state. A high K concentration therefore increases the efflux of K from cells into the tubular lumen and determines the rate of net secretion of K. Conversely a low K concentration reduces the efflux of ions from cells into the tubular lumen. It is at this stage that an alkalosis can exist with a paradoxical acid urine as the kidney has no alternative but to excrete H ions in exchange for Na. In this way tubular cells are simply behaving in the same way as the cells in the rest of the

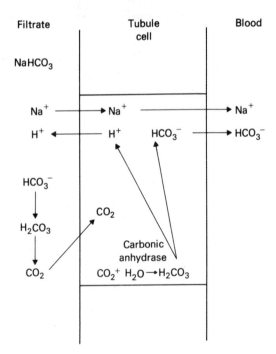

Fig. 3.6 Reabsorption of bicarbonate by the renal tubules.

body. Diminished Na intake is associated with reduced K excretion by a mechanism which is not fully elucidated.

Bicarbonate (HCO₃). This ion may be taken as an index of H ion secretion (Fig. 3.6). Sodium is reabsorbed into the blood in exchange for H ions which arise from carbonic acid (facilitated by carbonic anhydrase). Bicarbonate ions for reabsorption with Na ions are formed from carbon dioxide (CO_2). In K depletion, plasma HCO_3 is a quantitative measurement of the depletion. Plasma HCO_3 is raised due to accumulation of H ions within the K depleted cell. In metabolic alkalosis H ion secretion may be augmented due to the increased HCO_3 load delivered to the distal tubule. Hence, as discussed above both H and K secretions are greater than normal. In metabolic acidosis there is a reduction in the filtered HCO_3 load. Only small amounts of HCO_3 reach the distal tubule, most being reabsorbed in the proximal tubule and loop of Henle. Therefore less is available to permit H ion excretion so that H and K excretion are both reduced.

SODIUM POTASSIUM ADENOSINE TRIPHOSPHATASE

The activity of this enzyme is induced by K loading and is entirely independent of aldosterone. When the level of ATPase is raised isolated perfused kidneys can excrete up to three times the amount of K filtered at the glomerulus. This mechanism is important in maintaining K excretion per nephron as renal function is reduced.

FLOW RATE IN THE DISTAL TUBULE

In the distal renal tubule the K gradient across the wall is *independent* of the tubular flow rate. Therefore a high rate of flow results in an increase in total K excretion and a low flow rate in a reduced K excretion.

Reduction in tubular flow. In a situation of Na or water depletion tubular flow will be reduced and in addition proximal tubular reabsorption of Na will be more complete. Referring to Fig. 3.4 there is a reduced gradient of electronegativity down the distal tubule and less passive transport of K into the lumen. Aldosterone has little effect in these circumstances. If, however, Na intake is high aldosterone will promote K excretion because there is more Na in the urine.

SUMMARY OF RESPONSES TO A RISE IN POTASSIUM INTAKE AND PLASMA CONCENTRATION

Ingestion of a high K load results in the following:
1 Stimulation of insulin and glucagon secretion.
2 Stimulation of aldosterone production.
3 Stimulation of Na/K ATPase in the renal medulla.

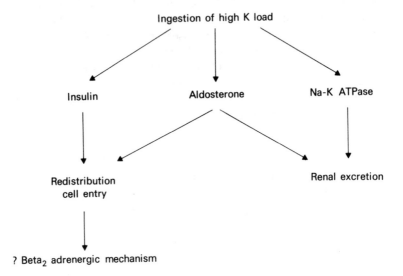

Fig. 3.7 Factors opposing hyperkalaemia after an acute potassium load.

Following a high oral K intake the K content of the small bowel lumen approaches plasma concentration rapidly and passive absorption occurs across the mucosa into ECF. However, ECF constitutes only a small percentage of the total body K pool so that unless ECF K is to reach a high level mechanisms must come into play rapidly to reduce this level.

If K is administered intravenously (i.v.) a more rapid rise in ECF K occurs since the stage of absorption across the mucosa is eliminated. This is why rapid i.v. administration of K may be particularly dangerous. In situations of chronically high K intake the efficiency of these mechanisms to limit the rise in plasma K is increased.

Sudden changes in K intake are rare, however, in humans. New Guineans migrating from a region of high to low K intake have developed hypokalaemic muscle paralysis within the first two months of the change. It seems then that there is considerable delay in modification of K excretion.

In any assessment of K balance it is important to remember that decreased K intake (effectively starvation) or an increased loss can to some extent be buffered by release of K from cells. Likewise increased K uptake by cells limits an acute rise in plasma K provided acid base regulation is normal, there is no lack of insulin and no excess catabolism present.

Chapter 4
Normal Calcium, Phosphate and Magnesium Metabolism

I'll grind these bones to make my bread. . . .

Jack and the Beanstalk

Calcium (Ca), phosphate (PO_4) and magnesium (Mg) are all essential for normal function in man. Transport and distribution of these three ions are often closely associated although their intra and extracellular concentrations are independently determined. Significant abnormalities may be present without any alteration from normal plasma concentration of Ca, PO_4 or Mg. This is especially the case in abnormal acid base states, chronic renal or gastrointestinal disease or with the ever increasing problem of drug induced enzyme induction. Similarly, an abnormal plasma level may be present without disturbance in total metabolism, for example rapid uptake of PO_4 into cells occurs when insulin stimulates glucose uptake resulting in a lowering of plasma PO_4 level.

CALCIUM

Ca is the fifth most abundant element in the human body. Most of the Ca within the body is found in the skeleton. Ca is required for the following:
1   Calcification of bones and teeth.
2   Regulation of metabolism in all cells and excitability of nerve cells.
3   Cardiac conduction.
4   Blood clotting.
In an average 70 kg man, Ca constitutes about 2% of the total body weight and amounts to 32 500 mmol (1000 g).
   99% of total body Ca is found within bone (32 125 mmol).
   0.5% is found in teeth (175 mmol).
   0.5% is found in soft tissues.

Small amounts are found in plasma (8 mmol) where almost half is bound to albumin so that hypoalbuminaemia produces a low plasma Ca level with no disturbance in total body Ca or turnover of Ca.

Normal plasma Ca is 2.2–2.6 mmol l^{-1} (2×10^{-3} M).

Normal urine Ca is 2.5–7.5 mmol daily.

Table 4.1   Constitutents of plasma Ca.

	Plasma concentration (mmol l^{-1})
Free Ca	1.18
Albumin bound Ca	1.14
Calcium phosphate ($CaHPO_4$)	0.04
Calcium citrate, pyruvate, lactate	0.04
Unidentified	0.08
	2.48

Intracellular Ca concentration is 10^{-7} mol l^{-1} and the large concentration gradient across the cell wall is maintained by Ca/Mg ATPase activity. Entry of Ca into cells occurs down the concentration gradient through Ca channels in the cell membrane.

It is important to sample blood for Ca measurement without use of a tourniquet because this encourages fluid loss from the vein and hence a spuriously high plasma protein concentration. It is quite possible to have a considerable abnormality in total body Ca with a normal plasma level. Ca binding to proteins is also affected by pH so that acute acidosis raises ionized Ca without affecting total Ca. The effects of metabolic acidosis are considerably greater than respiratory acidosis because pH not only determines the degree of interaction of Ca with proteins but also alters the degree of Ca-bicarbonate complex formation.

Methods of measurement of plasma calcium

1   Atomic absorption spectroscopy.

2   Chemical methods.

Most laboratories analyse Ca in an autoanalyser, for example an SMA. One method uses an indicator, orthocresol-phthalein. This indicator binds Ca at the correct pH. The degree of binding is proportional to the plasma Ca concentration and results in a colour change.

3 Ion selective electrodes. Only ionized Ca is biologically active. It is this component which is measured by these electrodes. However, this forms only half of total plasma Ca since the rest is protein bound.

Clinicians usually have to rely on measurements of total plasma Ca.

Correction of plasma calcium

It is possible to apply correction factors to take abnormal plasma proteins into account. Most of these corrections use the plasma albumin as their basis. A commonly used correction factor is as follows:

For every $1 \, g \, l^{-1}$ by which the plasma albumin exceeds or is lower than $40 \, gl^{-1}$ subtract or add $0.02 \, mmol \, l^{-1}$ to the measured plasma Ca. The total Ca therefore is corrected to an arbitrary normal plasma albumin and although inaccuracy may occur, for clinical purposes in a stable patient this correction prevents the clinician diagnosing hypocalcaemia in the hypoalbuminaemic patient or hypercalcaemia in a sample taken with excessive venous stasis. However, large changes in pH are not accounted for by these adjustments, Ca is also bound to globulins which may be important in diseases such as cirrhosis and myeloma and there is wide individual variation in the amount of Ca bound to albumin.

Many chemical pathologists, therefore, condemn such corrections pointing out that these formulae have been worked out over a limited range of plasma protein concentrations and therefore may not apply in extremes such as found in the intensive care unit. Different laboratory methods in the presence of an unknown acid base state can make blind reporting of corrected Ca hazardous and the best approach may be to record a straight measured plasma Ca pointing out any abnormality in plasma proteins.

There are now alternatives; measurements can be made of ultrafilterable, dialysable or ionized serum Ca. Ion selective electrodes, now much more reliable, should therefore be used if there is a high index of suspicion that total serum Ca may be misleading. Ion selective electrodes of course measure the activity of Ca in solution and this has to be multiplied by an activity coefficient dependent upon the ionic strength of the solution. In plasma this is determined by the Na and K concentrations and is fairly constant but ionized Ca also correlates with serum albumin (Donnan effect) so that Ca will be overestimated if it is measured in the presence of albumin. A further alternative is to measure dialysable Ca which is not interfered with by protein binding and only slightly influenced by posture although it still does not take into account large changes in anion concentration such as bicarbonate which will increase the complexed but not the ionized fraction of serum Ca.

Physiological importance of calcium

Bone

99% of total body Ca is found in bone as hydroxyapatite. This is largely found as the supporting structure of bone but small amounts of Ca in equilibrium with this large fixed amount are exchangeable with ECF and soft tissue Ca. Hence an enormous reservoir of Ca exists which through this small exchangeable pool can readily compensate for a fall in plasma Ca.

Cell membrane function

Ca is concerned with the following membrane functions:
1 Integrity.
2 Permeability.
3 Adhesiveness.
4 Intercellular connections.

Coupling of electrical or chemical excitation to intracellular contractile or secretory events

1 Specific Ca binding sites have been postulated on cell membranes which alter the permeability to Na and hence the excitability of the cell.
2 Ca is necessary for synaptic release of acetylcholine following an action potential in the nerve which then leads to muscle contraction.
3 Ca mediates the insulin secretion response to hyperglycaemia.

4 Close regulation of the plasma Ca contraction is necessary for normal muscle activity. A reduction in ionized Ca produces tetany whereas hypercalcaemia produces muscle weakness and paralysis. For initiation of muscle concentration, Ca moves from its binding site on the sarcoplasmic reticulum, via an ATP dependent mechanism, to actomyosin. Muscle contraction occurs which is terminated when Ca ions are released. It is well known from experiments in animal physiology that if Ca is absent from fluid perfusing the isolated heart it will stop beating. Addition of Ca restarts the heart. There is a significant correlation between low cardiac output and lowered ionized Ca levels.

All mitochondria can accumulate Ca by high affinity, active uptake mechanisms. The mechanism is ATP and oxidative phosphorylase dependent and serves to keep intracellular Ca levels low. In the mitochondria at least three dehydrogenases are activated by Ca and this is a very important means of enhancing oxidative phosphorylation.

Blood coagulation

Ca is important at several stages in the coagulation cascade.

Complement activation

This process proceeds in a similar cascade fashion to coagulation and Ca is again involved in this complex pathway.

If cell membrane function fails, the influx of Ca across the membrane along a steep concentration gradient, converts a potentially reversible alteration into irreversible cell death.

Calmodulin

Calmodulin is a Ca binding protein which has been intensively studied over the last 10 years. When intracellular Ca level rises from the normal 10^{-7} mol 1^{-1} following stimulation of the cell calmodulin binds Ca ions and the complex reacts with receptor protein.

Calmodulin has a MW 16 700 and four Ca binding sites of variable avidity. When Ca is so bound it has a helical structure and is biologically active. Some reactions require four bound ions, others only one.

The calmodulin calcium complex regulates a wide variety of essential cell processes but especially those involved in cyclic nucleotide and glycogen metabolism and smooth muscle contraction. Calmodulin-Ca and cAMP are inextricably linked and probably autoregulatory. Calmodulin activates both adenylate cyclase and phosphodiesterase which hydrolyse ATP and cAMP respectively and also activates the Ca pump which returns and raises intracellular Ca to resting level.

Calmodulin activates phosphorylase kinase in glycogenolysis and myosin light chain kinase in smooth muscle contraction. Others effects include those on platelet function, microtubular assembly, neurotransmitter release and enterocyte secretion. Some of the calcium channel blockers used in cardiovascular disease may act by inhibiting calmodulin and local anaesthetics inhibit calmodulin dependent phosphodiesterases. By a similar action phenothiazines may be useful antisecretory drugs in treating cholera. Enterocyte secretion is normally stimulated by Ca and induced by theophylline and diseases such as cholera.

Table 4.2   Enzymes and cellular processes regulated by calmodulin-Ca complex.

Adenylate cyclase
Phosphodiesterase
Phospholipase A_2
Glycogen synthase kinase
NAD kinase
Ca dependent protein kinase
Neurotransmitter release and mobilization
Ca-pumping ATPase (red cells)
          (sarcoplasmic reticulum)
Membrane phosphorylation
Microtubule assembly
Myosin light chain kinase
Phosphorylase B kinase
Guanylate cyclase

Normal calcium turnover

Dietary sources of Ca are plentiful. It is found abundantly in milk, cheese and eggs. Minimum requirements are of the order of 10 mmol daily. There is some evidence that the risk of colorectal cancer is inversely related to the dietary intake of vitamin D and calcium.

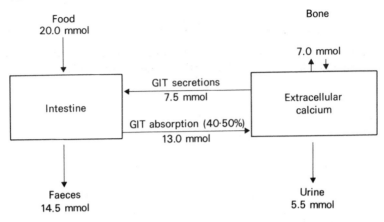

Fig. 4.1   Average daily calcium turnover for a 70 kg man. Ca MW = 40.08. GIT = gastrointestinal tract.

40–50% of dietary (and gastrointestinal secretion) Ca is absorbed in the small intestine. In extremes of intake this may range from 20–80%.

1   Active transport of Ca occurs in the duodenum, where Ca is absorbed against an electrochemical gradient. The active mechanism is located within the mucosal cells. Subsequent absorption into the blood may be passive.

2   Passive absorption of Ca also occurs. The amount of absorption of Ca depends on the body *requirements*, not on Ca supply. Therefore absorption is increased during active growth periods in early life and during pregnancy and lactation. At a given steady growth rate, on a low Ca diet, a greater proportion of the Ca intake will be absorbed; the amount depending on the amount of Ca intake and the level within the small intestine.

3   Ca secretion. This occurs in the small intestine relatively independently of the intake or growth rate and is therefore much less variable than absorption.

Ca intake varies widely with a mean of about 20 mmol from food. Faecal losses amount to about 14.5 mmol and urinary losses to about 5.5 mmol daily. Dietary intake of Ca has little effect on renal losses.

Factors affecting gastrointestinal absorption of calcium

Absorption is decreased by phosphates and oxalates as these form insoluble salts with Ca. Absorption is increased by the following factors:

1 Vitamin D.
2 Parathyroid hormone (PTH).
3 Growth hormone.
4 Corticosteroids.
5 Lactose.

Vitamin D

It is well known that vitamin D deficiency produces rickets in children and osteomalacia in adults. In animals a response to vitamin D administration is delayed for some hours, presumably whilst some vitamin D is converted into the active form (see below). The precise action of vitamin D in the intestine is not known but it may involve increased production of Ca binding protein which is known to be raised during periods of active growth and in conditions of low Ca intake. In addition vitamin D enhances the activity of Ca activated ATPase in the intestinal mucosa and this may be the mediator of vitamin D induced Ca transport.

PTH

In man hypocalcaemia stimulates PTH to increase the proportion of Ca intake which is absorbed BUT a decrease in the concentration of plasma Ca would normally improve passive Ca absorption because the gradient between intestinal lumen and plasma is more favourable.

Growth hormone

In growth hormone deficiency states Ca absorption is depressed.

Corticosteroid

Glucocorticoids depress the stimulatory effect of vitamin D on intestinal Ca absorption.

Lactose

Lactose increases intestinal absorption of Ca.

Approximately 13 mmol Ca daily are absorbed from the intestine into ECF from which there is interchange with Ca within bone during deposition or reabsorption (7 mmol daily) and loss of Ca into gastrointestinal secretions (7.5 mmol daily). Although adequate absorption of Ca and PO_4 from the gut is essential for mineralization of bone, the ECF and Ca content and hence plasma concentration is minimally affected by dietary intake or absorption, being closely regulated by hormonal mechanisms. Regulation of Ca metabolism is extremely complex and aimed at maintaining intracellular ionized Ca at a concentration of about 10^{-7} M. It is this concentration which is critically important for muscle activity, endocrine secretion and nervous system function and in the long term depends on maintenance of ECF Ca by hormonal activity.

Regulation of plasma calcium level

This hinges around three hormones: vitamin D, PTH and calcitonin. Other factors are of lesser importance.

Vitamin D

Historically vitamin D describes that constituent of the diet that prevents rickets. Cholecalciferol (D_3) occurs naturally and is made in the skin by the action of sunlight (ultraviolet) on the provitamin. Ergocalciferol (D_2) is present in fish liver and is derived from a provitamin found in some plants and fungi. Vitamin D_3 itself, however, is almost inactive.

There have been enormous developments in the last 10 years in recognizing the metabolites of vitamin D. Cholecalciferol is hydroxylated in the liver and perhaps other organs to produce 25 hydroxycholecalciferol ($25(OH)D_3$). Many enzyme inducing drugs may increase the microsomal process so that more polar, less active derivatives are produced that are excreted in the bowel. $25(OH)D_3$, however, is still relatively inactive although it is the major circulating form of the vitamin. It is transported on a specific gamma globulin binding protein to the kidney, where further hydroxylation occurs to produce several metabolites the most important of which are:

1   1,25 dihydroxycholecalciferol ($1,25(OH)_2D_3$)
2   24,25 dihydroxycholecalciferol ($24,25(OH)_2D_3$).

This second hydroxylation is crucial and was originally thought to occur only in the kidney.

$1,25(OH)_2D_3$

This is an extremely active metabolite with the following actions:

1   It stimulates the absorption of Ca and PO_4 from the gut.
2   It stimulates Ca and PO_4 resorption from bone.
3   It stimulates the reabsorption of Ca and PO_4 from renal tubules.

The overall effect of $1,25(OH)_2D_3$ is maintenance of Ca and PO_4 concentration in ECF and promotion of normal bone calcification. The healing action of vitamin D in rickets seems to be entirely due to its action in raising plasma Ca and PO_4 levels.

$1,25(OH)_2D_3$ levels are high in conditions of increased demand such as rapid growth, pregnancy and lactation, and may be regulated by the pituitary. Low concentrations of plasma Ca or PO_4 increase circulating $1,25(OH)_2D_3$ via a feedback loop.

$24,25(OH)_2D_3$

This is definitely a less active metabolite which may promote normal formation and mineralization of bone or represent simply a diversionary pathway used when there is no increase in Ca requirement. $1,25(OH)_2D_3$ is controlled by three regulators;

(a)   Factors involved in the regulation of serum Ca
(b)   Factors involved in increasing $1,25(OH)_2D_3$ in response to the increased physiological requirements of growth and pregnancy (growth hormone, prolactin, calcitonin, placental lactogen, oestradiol).
(c)   Factors involved in maintaining vitamin D activity in the face of variable sunlight exposure.

Parathormone

PTH only evolved with land mammals when phosphate intake became important. The phosphaturic effect of PTH allowed survival on land with a high PO_4 diet. PTH is a polypeptide (MW

9500) consisting of a single chain of 84 amino acids. It has long been held that when plasma levels of Ca, Mg and other divalent ions fall, secretion of this hormone occurs from the chief cells of the parathyroid glands. Plasma levels of Ca and PO_4 are then restored towards normal by an action on the bone and kidneys. Secretion of PTH is contolled mainly by circulating Ca and Mg ions. Mg is necessary both for parathyroid cell secretion and PTH receptor activity. $1,25(OH)_2D_3$ suppresses PTH secretion probably by facilitating transport of Ca into parathyroid cells.

BONE

Within bone both osteoblastic and osteoclastic activity are stimulated resulting in release of Ca and PO_4 into ECF and therefore increasing plasma levels. This process involves $1,25(OH)_2D_3$.

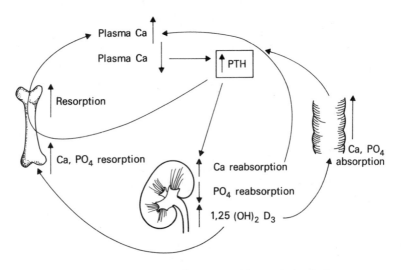

Fig. 4.2 The relation between parathormone, calcium and vitamin D.

KIDNEY

PTH increases renal tubular reabsorption of Ca and Mg and reduces that of PO_4.

The postulated mechanism for regulation of plasma Ca therefore is that hypocalcaemia stimulates PTH production which increases $1,25(OH)_2D_3$, mobilizes Ca and PO_4 from bone, increases gastrointestinal absorption of Ca and PO_4 and permits PO_4 loss with Ca retention by the kidney. As the plasma Ca is restored, $1,25(OH)_2D_3$ levels return to normal and $24,24(OH)_2D_3$ levels increase. Until recently it was thought that PTH was essential for regulation of vitamin D metabolism but it now appears that increased levels of $1,25(OH)_2D_3$ do not depend primarily on variation of PTH secretion although considerable controversy still exists regarding the regulation of Ca metabolism by vitamin D and PTH.

PTH has no direct action on gut but may indirectly lead to increased intestinal absorption of Ca by stimulating $1,25(OH)_2D_3$ production.

Calcitonin

This hormone is a 32 amino acid polypeptide from the parafollicular (C) cells of the thyroid which is released when ECF Ca levels rise. C cells also occur in the thymus, especially in early life, and in the lung. Its actions reduce plasma Ca.

ACTION OF CALCITONIN

1 Inhibition of bone resorption. This is the main effect of calcitonin.
2 Renal excretion of Ca and PO_4 are increased whereas that of Mg is decreased.
3 Gastrointestinal absorption of Ca and PO_4 may also be reduced.

Inhibition of skeletal resorption results in a fall in urine hydroxyproline. If calcitonin is injected into normal adults there is little or no change in plasma Ca. This is because under normal circumstances the Ca efflux from bone forms only a small percentage of the net movement of Ca into plasma. Calcitonin will produce hypocalcaemia, however, during periods of active growth or in Paget's disease where bone turnover is enhanced. There is normally marked diurnal variation in plasma levels of calcitonin.

FACTORS WHICH STIMULATE CALCITONIN SECRETION

1 Hypercalcaemia.
2 Glucagon.
3 Pentagastrin.
4 Cholecystokinin.
5 Catecholamines.
6 Alcohol.
7 Presence of food in stomach.
8 Oral contraceptive pill.
9 Pregnancy.
$1,25(OH)_2D_3$ may increase calcitonin secretion probably by promoting Ca ion transport into cells.

Under normal physiological conditions calcitonin is of little importance. It does have a role in regulating plasma Ca but may be more important for ensuring skeletal homeostasis and it may be

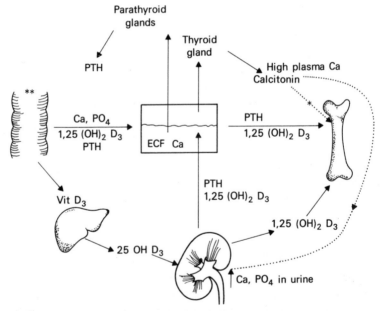

Fig. 4.3 Sites at which vitamin D, parathormone and calcitonin affect plasma calcium concentration. *$1,25(OH)_2D_3$ may also stimulate deposition of Ca, PO_4 in bone. **Calcitonin may reduce gastrointestinal absorption of calcium and phosphate.

that lack of calcitonin is an important factor in the rapid bone loss which occurs in women after the menopause. There is a striking sex difference in circulating plasma calcitonin with a relative deficiency in women. Calcitonin secretion in young adults is increased by oestrogens and recent studies of the effects of oestrogen in postmenopausal women show a sharp increase in plasma calcitonin. Oestrogens prevent menopausal bone loss and this effect could be mediated at least in part by control of calcitonin secretion.

THYROXIN AND TRIIODOTHYRONIN

These hormones increase bone Ca and PO_4 turnover. Hypercalcaemia may be seen in thyrotoxicosis.

GROWTH HORMONE

In addition to its effect on intestinal absorption, growth hormone also induces retention of Ca in bone.

GLUCAGON

Glucagon may produce a fall in plasma Ca by either direct suppression of bone resorption or by release of calcitonin. Glucagon release may be one cause of hypocalcaemia in pancreatitis.

GLUCOCORTICOIDS

It is well known that chronic steroid therapy leads to osteoporosis. In addition to their effect on the gastrointestinal tract, glucocorticoids inhibit bone resorption and promote Ca and PO_4 excretion in the urine.

Summary of interactions of calcium regulating hormones (Fig. 4.3)

1 The gastrointestinal tract, kidney and bone effect Ca homeostasis.

2 $1,25(OH)_2D_3$ is the main regulator of Ca and PO_4 absorption from the gut and is important mainly for providing Ca and PO_4 for the fetus and growing child. Its destructive effect on bone is normally prevented by calcitonin since levels of both these

hormones are increased in growth, pregnancy and lactation. Therefore, the presence of calcitonin is necessary for the normal action of $1,25(OH)_2D_3$.

3 PTH controls Ca excretion by the kidney. It promotes excretion by PO_4 by the kidney whilst enhancing tubular reabsorption of Ca. Its less important action is on the gut and is indirectly enhancing Ca and PO_4 absorption by an increase in $1,25(OH)_2D_3$ production. Direct action of PTH on bone is probably not important in maintaining plasma Ca.

4 Calcitonin controls the rate of skeletal resorption by interaction with $1,25(OH)_2D_3$.

Calcium and the kidney

Renal handling of Ca is similar to that for Mg (see below) except that both free and some unionized complexes of Ca are filtered at the glomerulus and there is no doubt that PTH specifically stimulates resorption of Ca. A reciprocal relationship exists between Ca and PO_4 excretion. Ca and Mg are maximally excreted in the morning, PO_4 in the afternoon.

In addition to direct regulation of Ca, Mg and PO_4 by hormones and cellular transport mechanisms there is indirect regulation due to interaction with other ions. For example in metabolic alkalosis (low hydrogen (H) ion concentration) there is increased muscular excitability, as in tetany due to hypocalcaemia. This is only partly due to an alteration in the relationship between free and bound Ca, and effects of H ion concentration on Ca binding at nerve membranes may also be important. In addition reciprocal movement of Ca and H ions can be observed in mitochondrial transport.

It must be stressed that it is the free ionized Ca which is taking part in essential physiological mechanisms but that this constitutes only about 50% of the total plasma Ca. The extent of Ca binding by plasma proteins is in direct proportion to the plasma albumin level, therefore it is essential for this to be measured at the same time as the Ca measurement is undertaken.

PHOSPHATE

Phosphorus constitutes 1% of the total body weight and is the sixth most abundant element in the body. Total body phosphorus content is 22 600 mmol. Within the body phosphorus is in the form of phosphate ion, PO_4, and it is this form which is measured in the plasma.

Table 4.3 Distribution of phosphorus in a 70 kg adult man.

	%	mmol
Bone	85.0	19 200
Teeth	0.4	98
Soft Tissue	14.3	3230
Extravascular fluid	0.03	7
Blood		
plasma	0.02	4
erythrocytes	0.3	61

Normal plasma PO_4 concentrations is 0.8–1.35 mmol l^{-1}.
Normal urine PO_4 excretion is 4.0–17 mmol l^{-1}.

As with Ca, significant derangement can occur without a significant abnormality in plasma PO_4 level. This is especially so in the following circumstances:
1 Acid base disturbances.
2 Chronic renal and gastrointestinal disease.
3 Drug induced abnormalities of vitamin D metabolism.
Equally either high or low plasma PO_4 levels may occur without significant disturbances of total body PO_4 content or turnover for example during haemoconcentration or dilution, during haemolysis and during hydrolysis of PO_4 containing compounds.

Measurement of plasma PO_4 levels

Blood should be sampled from a fasting patient without use of a tourniquet and the laboratory should be informed if a glucose infusion is in progress as this lowers plasma PO_4 owing to uptake of PO_4 into the cells with glucose under the influence of insulin.

Measurement of plasma PO_4 utilizes a chemical method first described by Fisk and Subbarow. It involves a reduction of phosphomolybdate to molybdenum, when a colour change takes place proportional to the PO_4 concentration. PO_4 may be routinely measured in an autoanalyser. It is present in several forms within plasma.

Table 4.4 Plasma PO_4 concentration (mmol 1^{-1}).

Free HPO_4^{--}	0.5
Free $H_2PO_4^-$	0.11
Protein bound	0.14
Na_2HPO_4	0.33
$CaHPO_4$	0.04
$MgHPO_4$	0.03

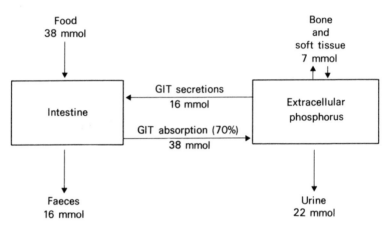

Fig. 4.4 Daily turnover of phosphorus. MW = 30.98; GIT = gastrointestinal tract.

Daily turnover of phosphorus

Dietary PO_4 is in good supply. Minimum daily requirements are 25 mmol. The average daily adult intake is about 40 mmol. In the adult only enough PO_4 is retained to compensate for normal losses via the urine and faeces but in children during growth, or in pregnancy and lactation requirements are increased with net PO_4 retention. Almost three quarters of dietary PO_4 is absorbed with faecal losses amounting to 14–16 mmol daily. Low dietary intake of PO_4 reduces plasma PO_4 in normal people and high intake results in increased urinary excretion. Plasma PO_4 levels may be markedly reduced during insulin facilitated uptake of glucose into cells. Once absorbed into ECF some 7 mmol PO_4 interchanges daily with bone PO_4. 16 mmol daily appears in gastrointestinal tract secretions and 22–26 mmol is excreted in the urine.

Physiological importance of PO_4

85% of PO_4 is located in bone and 15% in soft tissues within cells as organic phosphorus compounds.

Intracellular phosphate

Intracellular phosphate occurs in the following forms:
1 Adenosine triphosphate (ATP). This nucleotide is a labile high energy PO_4 compound, which is the final common energy providing compound for many reactions such as muscle contraction, hormone secretion and active transport of substances across cell membranes.
2 Nucleic acid.
3 Phospholipids.
4 Phosphoproteins.
Many of these compounds are important in maintaining the structure and function of cell membranes.

Buffer systems

Phosphate compounds exist as acidic or basic salts ($H_2PO_4^-$, HPO_4^{--}) and hence they form buffer systems in urine and cells.

Hypophosphataemia

Hypophosphataemia is now becoming increasingly recognized during parenteral nutrition when high concentrations of glucose containing solutions cause glucose and phosphate to enter the cells under the influence of insulin. Plasma PO_4 is an important determinant of red blood cell 2,3 diphosphoglycerate (2,3 DPG) which in turn determines the position of the oxygen dissociation curve. A reduction in 2,3 DPG results in a left shift of the oxygen dissociation curve and impaired delivery of oxygen to the tissues, with consequent hypoxia. It must be remembered, however, that hypophosphataemia is almost routine postoperatively but at a level which causes little reduction in 2,3 DPG. Hyperphosphataemia on the other hand increases red blood cell 2,3 DPG.

Regulation of phosphate turnover

This has largely been covered under the section on Ca. Essential hormones for control of PO_4 metabolism are:

1 PTH.
2 $1,25(OH)_2D_3$.
3 Calcitonin.

Less important hormones concerned with PO_4 metabolism are growth hormone, prolactin, corticosteroids and thyroxin. PTH acts on specific cell membrane receptors linked to adenyl cyclase in kidney and bone. It releases cyclic adenosine monophosphate (cAMP) which transmits the stimulus within the cell. In the absence of PTH 90% of PO_4 is reabsorbed in the renal tubule. PTH reduces tubular reabsorption of PO_4, lowers the renal threshold and allows some excretion of PO_4 at lower plasma concentrations. However, PTH also stimulates the formation of $1,25(OH)_2D_3$ from $25(OH)D_3$. $1,25(OH)_2D_3$ is responsible for increasing the plasma PO_4 concentration mainly by increasing intestinal absorption. An increase in plasma PO_4 inhibits synthesis of this hormone by negative feedback.

CALCITONIN

The same comments apply as for the section on Ca. Calcitonin also acts via adenyl cyclase on the kidney and has a generalized effect, reducing tubular absorption of Ca, PO_4, Na, Mg, K and HCO_3.

There is little evidence to show that plasma PO_4 requires such stringent regulation within limits as does Ca in fact considerable fluctuations occur in plasma PO_4 throughout the day. There are no specific direct mechanisms to control plasma PO_4. It is postulated, however, that low concentrations of intracellular inorganic phosphorus stimulate the formation of $1,25(OH)_2D_3$ with subsequent increase in Ca and PO_4 absorption from the gut. If ECF Ca is normal this will have the effect of suppressing PTH so the kidney will excrete Ca but PO_4 reabsorption occurs. Impaired renal function has a marked effect on plasma PO_4. A reduction in glomerular filtration rate (GFR) reduces excretion of PO_4 with hyperphosphataemia. If tubular function is impaired PO_4 reabsorption is reduced producing hypophosphataemia and PO_4 depletion.

Thyroxine increases bone Ca and PO_4 turnover. Growth hormone increases intestinal absorption of PO_4 and reduces renal tubular reabsorption of PO_4. This is reflected in a small positive PO_4 balance during growth. Active acromegaly may be accompanied by hyperphosphataemia. Glucocorticoids increase PO_4 excretion in the urine.

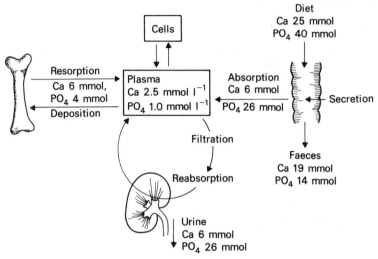

.**Fig. 4.5** Summary of daily calcium and phosphate turnover showing approximate net values.

MAGNESIUM

Magnesium (Mg) is the eleventh most abundant element in the body, the fourth most abundant cation and the second most abundant intracellular cation after K. 54% of the total Mg is found in

Table 4.5 Distribution of magnesium in a 70 kg man (mmol).

Bone	610
Soft tissue (muscle)	500
Extravascular fluid	7
Blood	
plasma	2.5
erythrocytes	5.5
	1125

bone. Soft tissue Mg is largely intracellular and mainly in the bound form. Normal plasma Mg is 0.7–1.0 mmol l^{-1}; about one third being in the protein bound form.

Plasma Mg within the normal range is not uncommon with a total body Mg depletion which is similar to the situation with K. (Only 1% Mg is extracellular.)

Normal urine Mg is 2.5–5.0 mmol l^{-1}.

Table 4.6 Constitution of plasma magnesium (mmol l^{-1}).

Free Mg	0.53
Protein bound	0.30
$MgHPO_4$	0.03
Mg citrate	0.04
Other complexes	0.06

Only 15% of Mg exchanges rapidly and most of this is in soft tissues. Bone does not act as a reservoir in the same way as it does for Ca. Calcium and Mg ions probably compete for albumin binding sites but only 10% of the sites are occupied by Mg. Intracellular Mg in most tissues is about 10 mmol l^{-1}, mainly bound to nucleic acids and proteins (0.1–1 mmol l^{-1} is free).

Measurement of plasma magnesium

Two methods are available which have been discussed elsewhere:
1 Atomic absorption.
2 Flame spectroscopy. This was the method used prior to the development of atomic absorption spectroscopy.

Daily magnesium turnover

The average daily Mg intake is about 12 mmol largely from green vegetables, cereals, nuts and dairy produce. Minimum requirements have been put at 3 mmol daily. Such a value is a result of studies of the minimum intake required to prevent negative Mg balance.

Less than half the ingested Mg is absorbed although this can rise to 80% on low Mg intakes. Faecal losses amount to 8 mmol daily and urine losses to 4 mmol daily although this can be reduced during deprivation to less than 0.5 mmol daily.

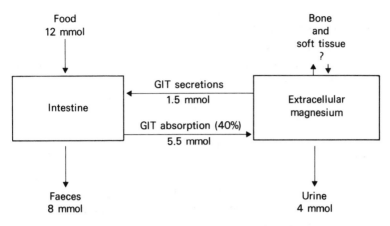

Fig. 4.6 Daily magnesium turnover. Mg MW = 24.32. GIT = gastrointestinal tract.

About 40–50% of Mg is absorbed from the gut into ECF from which some is secreted in the urine, some is deposited in bone and about 1.5 mmol daily enters gastrointestinal tract secretions. Dietary intake and absorption do influence plasma concentrations of Mg but have little effect on renal excretion. Little is known about the intestinal absorption of Mg. There may be a common mechanism for Mg and Ca absorption since in animal nutrition studies there is competition between Mg and Ca for absorption. Intestinal absorption is probably increased by both vitamin D and PTH and reduced by phosphate binding. There is evidence for a transport system under genetic control specific for Mg without which a normal plasma Mg cannot be maintained on normal dietary intake.

As glucose enters cells stimulated by insulin K, PO_4 and Mg enter as well. It is well known that intravenous infusions of glucose, sorbitol and other carbohydrates may be associated with hypomagnesaemia as well as hypophosphataemia.

Physiological importance of magnesium

Within cells Mg is specifically associated with the function of a number of intracellular structures and organelles.

Enzyme systems

The most clearly defined action of Mg is to activate intracellular enzyme systems. That Mg is found abundantly in pre-Cambrian seas points to the evolutionary importance of this.

ENZYME SYSTEMS ACTIVATED BY Mg

1 All ATPases, alkaline phosphatases and pyrophosphatases. Since ATP is required for all anabolic processes including fat, protein and nucleic acid synthesis and glucose utilization, Mg is essential for all these functions. There is a good correlation between high metabolic activity within cells and their Mg, K and high energy phosphate content.
2 Thiamine pyrophosphate as cofactor.
3 Enolase.
4 Some peptidases.

Neuromuscular activity

Mg has properties similar to those of Ca. A fall in ECF Mg results in paraesthesia, increased neuromuscular excitability, tetany and cramps. A rise in ECF Mg is associated with muscle weakness, paralysis and cardiorespiratory collapse. Plasma Mg of more than 2.5 mmol l^{-1} inhibits atrioventricular and intraventricular conduction and 8–10 mmol l^{-1} produces cardiac arrest. Direct application of Mg to central nervous tissue blocks synaptic transmission and at high doses Mg salts produce loss of consciousness. The neuromuscular transmission block is associated with a reduction in amplitude of the end plate potential. Both central and peripheral depression of transmission can be antagonized by Ca.

Fish and invertebrates can be anaesthetized with Mg and returned to consciousness after intravenous Ca. These opposing effects are interesting in view of the similar effect of the two ions on normal excitability; low levels of either producing tetany.

Vascular system

Mg is a peripheral vasodilator which in large doses lowers the blood pressure.

Alterations of intra- and extracellular Mg concentration may affect cell function through effects on Ca handling. Competitive binding to some sites may produce an appropriate physiological response, or be without effect or displace Ca with increased effect. A high serum Mg suppresses PTH which is an example of Mg mimicking the effect of Ca. Eclampsia is often treated with Mg infusions which may act by redistribution of Ca and therefore reduction in blood pressure. Conversely Mg depletion may be associated with hypertension.

Normal regulation of magnesium

The plasma concentration of Mg is by no means as well controlled as that of Ca and the mechanisms are less well clarified. During Mg depletion plasma concentration can fall by 50% with very few symptoms and a relatively small loss of total Mg.

Factors concerned with normal regulation of magnesium

1 Hormones.
2 Ion balance across cell membranes.
3 The kidney.

HORMONES

Thyroid hormone. Excess thyroid hormone can decrease plasma Mg concentration and is associated with negative Mg balance.
Growth hormone. This causes retention of Mg in bone and soft tissues.
Parathormone. Only PTH seems to be normally concerned with homeostasis.
Aldosterone and vasopressin can influence Mg handling.

As plasma concentration of Mg falls the release of PTH is increased resulting in:
1 Diminished urinary Mg secretion.
2 Increased gastrointestinal absorption of Mg.
However, PTH is relatively inefficient at restoring plasma Mg to normal levels probably due to the absence of an efficient feedback mechanism. Calcitonin, corticosteroids, vitamin D and thyroid hormone all have a small indirect effect.

ION BALANCE ACROSS CELL MEMBRANES

As Mg is actively taken up into cells the intracellular concentration is maintained at the expense of the ECF concentration. In addition to the well known Na pump there may be a pump to regulate intracellular Mg, Ca ratios. Either Ca is pumped out of the cell or Mg is pumped in or both. Ca often functions as an extracellular activator whereas K and Mg activate intracellular phosphate metabolism.

RENAL EXCRETION

Free Mg is filtered at the glomerulus in an amount which varies with the glomerular filtration rate and the amount bound to plasma protein. Some of this is reabsorbed in the proximal tubule where PTH stimulates the process. Normally 3–8% of the filtered load is excreted in the urine. Proximal tubular reabsorption of filtered Ca and Mg is closely related to reabsorption of Na although in conditions of Mg deficiency the excretion of Mg is markedly reduced whilst that of Na remains the same. Mg reabsorption in the proximal tubule is not significantly affected by conditions such as acidosis, diuresis or fluid overload. The most important site of renal Mg manipulation is the loop of Henle where 50–60% of filtered Mg is reabsorbed and up to 80% in deficiency states. Reabsorption is probably related to active reabsorption of chloride creating a voltage gradient across which Mg travels passively. Active reabsorption of Mg also occurs in the distal tubule and collecting duct against an electrochemical gradient. There may be additional transport mechanisms shared with Ca since increasing the plasma concentration of one ion decreases reabsorption and increases excretion of the other ion. PTH enhances renal tubular reabsorption of Mg.

Chapter 5
Normal Acid Base Balance

White shall not neutralize the black, nor good
compensate bad . . .

The Ring and the Book

R. BROWNING

In a neutral solution there are equal numbers of hydrogen (H) ions and hydroxyl (OH) ions. An acid solution contains an excess of H and an alkaline solution an excess of OH ions.

pH is the negative logarithm to the base 10 of the H ion concentration. The pH scale indicates the number of H ions logarithmically.

Pure distilled water contains $0.000\ 0001\ mol\ l^{-1}$ ions and the same number of OH ions. Therefore the log to the base 10 of $0.000\ 0001$ being -7 the product of the H and OH ions in pure water is -14. In the pH range 0–14 an increase in H ion concentration and increased acidity results in a falling pH.

This scale is logarithmic so that each unit change in pH is equivalent to a 10-fold change in H ion concentration.

(pH)

	$(nmol^{-1})$
6.8 is equivalent to a H ion concentration of	160
7.1	80
7.4	40
7.7	20

(1 Nanomole $= 10^{-9}$ mole.)

The normal pH range of the body is 7.35–7.45. This is slightly alkaline.

Normal H ion concentration is 35–45 nmol l^{-1}.

An acid is a proton donor or H ion donor.

A base is a proton acceptor or H ion acceptor.

Acidaemia occurs when arterial blood pH is less than 7.36 or H ion concentration is greater than 44 nmol l^{-1}.

Alkalaemia occurs when pH is greater than 7.44 or H ion concentration is less than 36 nmol l^{-1}.

Acidosis and alkalosis are abnormal conditions which cause acidaemia and alkalaemia respectively *if* no secondary changes occur to compensate for the primary change. Although there are wide variations in daily acid and base intake there is no specific centre for H ion regulation.

Most physiological processes involve enzymatic activity which is optimal around the H ion concentration of body fluids. This influence is much more important than temperature or concentration of substrates. Acidosis also affects the degree of dissociation of drugs and therefore their lipid solubility and effectiveness; for example at low pH the potency of narcotic analgesics and narcotic antagonists is markedly reduced.

The fundamental problem for humans lies in the fact that cells produce large amounts of H ions during metabolism. Normal adult intermediary metabolism produces 50–100 mEq acid daily (0.8 mEq kg^{-1} day^{-1}) as H_2SO_4, H_3PO_4 and organic acids.

Table 5.1 Daily hydrogen ion balance (mmol day^{-1}).

Input		Output	
Volatile			
CO_2	13 000	Lungs	13 000
Lactate	1500	Liver/kidney	1500
Non-volatile			
Protein, SO_4, PO_4	45	Titratable acid	30
Phospholipid	13	NH_4	40
Other	12		

Metabolism of carbohydrate, fat and protein results in CO_2 production. Average O_2 consumption by the body is about 250 ml min^{-1} and CO_2 production 200 ml min^{-1}. The ratio between these two is the respiratory quotient;

$$\frac{CO_2 \text{ production}}{O_2 \text{ consumption}} = 0.8$$

During oxidation of glucose only;

$$C_6H_{12}O_6 + 6O_2 = 6CO_2 + 6H_2O$$

Glucose   Oxygen

(RQ = 1)

The CO_2 then reacts with water to form H ions

$$CO_2 + H_2O \rightleftharpoons H_2CO_3 \rightleftharpoons H^+ + HCO_3$$

Despite this reaction the H ion concentration is normally kept within strict limits by two mechanisms:

1 Buffering of the H ions.
2 Elimination of the H ions.

BUFFERING

The law of mass action states that in any chemical reaction at equilibrium $HA \rightleftharpoons H^+ + A^-$ where A is any anion and HA is undissociated acid. This may be written in an alternative form where K is the dissociation constant of the reaction and [] is concentration.

$$K = \frac{[H] + [A]}{[HA]} \text{ or } [H] = \frac{K[HA]}{[A]}$$

The negative logarithm of this equation gives the Henderson Hasselbalch equation.

$$pH = pK + \log \frac{[A]}{[HA]} \tag{1}$$

The combination of a weak acid and a strong base or a strong acid and a weak base constitutes a buffer solution into which addition of H or OH ions has little effect on the pH of the solution. In plasma and tissue fluid three buffer systems exist.

1 Proteins. $H^+ + Prot^- \rightleftharpoons HProt$
2 Phosphate. $H^+ + HPO_4^{--} \rightleftharpoons H_2PO_4^-$
3 Carbonic acid, bicarbonate. $H^+ + HCO_3^- \rightleftharpoons H_2CO_3 \rightleftharpoons CO_2 + H_2O$.

Within red blood cells haemoglobin forms an additional important buffering system. This is discussed below.

Each buffer system has a pK value at which its buffering capacity is maximal. K is the dissociation constant of the reaction and pK is the negative logarithm to the base 10 of the dissociation constant. Most buffering occurs within ± 1 pH unit of the pK value of the system.

Figure 5.1 shows the curve for the bicarbonate system which is most effective at pK 6.2. At plasma pH 7.4 therefore this system is relatively ineffective.

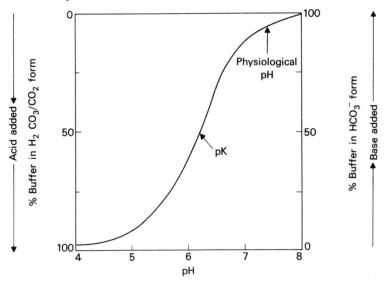

Fig. 5.1 The reaction curve for a buffer system.

Blood buffering systems

Table 5.2 Constituents of total blood buffering capacity.

	% Total buffering capacity
Plasma	
Bicarbonate	65
Phosphate	1
Protein	5
Erythrocytes	
Haemoglobin	29

Bicarbonate carbonic acid

This is the main buffer system in blood. The relation between bicarbonate and carbonic acid concentration is 20:1. Its pK value is 6.1 and therefore its chemical buffering capacity at pH 7.4 is poor but its efficiency increases when blood pH decreases. This is an open ended system because CO_2 can be adjusted immediately by ventilation in normal lungs and HCO_3 by the kidney in the longer term. The large amount of plasma HCO_3 available (24–28 mmol l^{-1}) makes this system especially important. Addition of alkali to this system will increase HCO_3 and eventually HCO_3 will be excreted in the urine.

Phosphate

This system has a pK of 6.8 and is therefore a better chemical buffer than the bicarbonate system but its concentration in plasma is much lower and therefore its capacity as a buffer is much less although it is important in urine and ICF.

Haemoglobin and protein

All proteins have a number of titratable groups within their molecular structure with ability to buffer pH changes. In haemoglobin this is due to the imidazole group of histidine which dissociates less in oxygenated than deoxygenated blood. Haemoglobin has three times the buffering capacity of plasma proteins gram for gram and twice the concentration therefore it has six times the total buffering capacity of plasma proteins.

Ionic shifts also occur between ICF and ECF during buffering. During respiratory acidosis for example both Na and K leave ICF for ECF and Cl moves inside the cell. 37% of the acid load is handled by Na proton exchange, 14% by K proton exchange and 29% by Cl bicarbonate exchange. In metabolic disturbances up to 50% of acid load is buffered intracellularly. Respiratory changes are predominantly buffered in the intracellular space as CO_2 changes occur in total body water.

In summary buffering is the mechanism by which an influx of H ions is initially dealt with by the body which limits the change in pH. Thereafter respiratory mechanisms become active to eliminate CO_2 and later renal regulation of acid secretion restores buffer capacity. If the plasma proteins or HCO_3 concentration are decreased the buffer capacity of the plasma will be reduced and an acid load will raise the H ion concentration more than usual.

Bone plays an important part in buffering. Chronic acid loading increases calcium excretion from bone and in general metabolic acidosis does decrease bone mass.

ELIMINATION OF HYDROGEN ION

Rapid elimination, respiratory response

Oxidation of energy substrates produces CO_2 which can be eliminated by the lungs. Arterial CO_2 varies directly with CO_2 production and inversely with alveolar ventilation. CO_2 is very soluble and readily diffuses across cell membranes producing changes in pH in both ICF and ECF. In the normal situation increased CO_2 production stimulates respiration and can be matched by increased alveolar elimination. Total ventilatory failure for 20 minutes produces a profound acidosis with an arterial PCO_2 of 14.7 kPa (110 mmHg) and pH 7.03 whereas renal failure for this length of time has no effect. The inefficiency of the carbonic acid bicarbonate chemical buffer system has already been discussed but carbonic acid is in equilibrium with dissolved CO_2 in body fluids and can therefore be eliminated through the lungs. Addition of H ions increases H_2CO_3 at the expense of a reduction in bicarbonate.

$$HCO_3^- + H^+ \rightleftharpoons H_2CO_3$$

more CO_2 is then formed;

$$H_2CO_3 \rightleftharpoons CO_2 + H_2O$$

CO_2 may be carried by the blood as physically dissolved CO_2, as carbonic acid, bicarbonate ion, carbamino-Hb and a very small amount as carbonate ion.

Slow elimination, renal response

Long term control of H ion excretion depends on renal mechanisms. Excess H ions produced by the body are dealt with by the following mechanisms:
1 Buffers.
2 Formation of ammonia.
3 Reabsorption of filtered bicarbonate.

Buffers

20–30 mmol H ions are excreted daily by conversion of monohydrogen to dihydrogen phosphate.

$$H^+ + HPO_4^- \rightleftharpoons H_2PO_4$$

Formation of ammonia

NH_3 is formed in the tubular epithelium throughout the nephron. 60% is formed from glutamine by deamination and deamidation, 30–35% from free arterial NH_3. This NH_3 diffuses into the renal tubular lumen and binds an H ion to produce non-diffusible ammonium (NH_4^+) which is excreted. In this way 30–50 mmol H ions are excreted daily under normal circumstances and this may rise to 700 mmol daily in severe acidosis.

Reabsorption of filtered bicarbonate

85–90% of filtered HCO_3 is reabsorbed in the proximal tubule. The amount of HCO_3 reaching the distal nephron varies with the filtered load of HCO_3 and effective ECF volume. Microperfusion

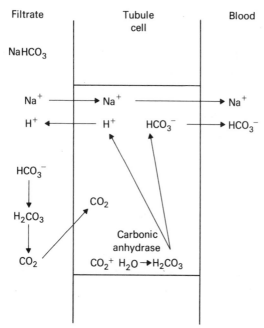

Fig. 5.2 Hydrogen ion excretion by the renal tubule.

techniques have shown that active and passive proximal acid base regulation depends upon luminal bicarbonate concentration, luminal flow rate, peritubular bicarbonate, PCO_2, ECF volume and solute solvent interactions.

In the proximal tubule, reabsorption occurs with H ion excretion. Carbonic anhydrase within renal tubular cells catalyses the hydration of CO_2 to carbonic acid which then ionizes to H and HCO_3 ions. Reabsorption of HCO_3 occurs into the blood stream and H ions are excreted in the urine in exchange for Na ions. The increase in HCO_3 reabsorption by the distal renal tubule in acidosis is progressive and takes five days to reach maximal response when 700 mmol H ion can be excreted daily. The raw materials for generation of HCO_3 by the kidney, H_2O and CO_2, are present in inexhaustible supply. In the absence of carbonic anhydrase, for example when the diuretic acetazolamide (a carbonic anhydrase inhibitor) is given, acid excretion is reduced, the urine becomes alkaline, large amounts of $NaHCO_3$ are excreted and plasma HCO_3 falls. Patients lacking carbonic anhydrase in red cells exhibit renal tubular acidosis, osteopetrosis and mental retardation.

Factors affecting proximal tubular H ion excretion

1 Intracellular acid base status which in turn varies with arterial CO_2, metabolic H and K ion status.
2 Luminal pH.
3 Functional ECF volumes.
4 Availability of reabsorbable anions.
5 Carbonic anhydrase.
6 Parathormone.

Hypokalaemia enhances proximal tubular H ion secretion by increasing the availability of H ion to the secretory mechanism (see Chapter 3). If effective ECF volume is depleted the Na:H ion exchange mechanism permits retention of Na so that ECF volume is maintained at the expense of pH regulation.

Factors affecting distal tubular H ion excretion

1 Intracellular acid base state.
2 Luminal pH.

3 Mineralocorticoid activity and K status.

Hyperaldosteronism enhances distal Na reabsorption and H and K ion excretion.

In clinical practice the only effective measurement that can be made is that of ECF status. This is assessed from arterial blood gas analysis. In the acute situation, however, there is a very poor correlation between this and the intracellular pH. This is the case in hypokalaemia where an ECF alkalosis exists with an ICF acidosis. In K depletion H ions enter the cell to maintain electrical balance and replace the intracellular K.

MEASUREMENT OF ACID BASE BALANCE

A heparinized arterial blood sample is taken to measure pH, $PaCO_2$ and HCO_3. The combined information is essential in the assessment of respiratory and metabolic disorders of acid base status.

Normal values

pH 7.36–7.44
$PaCO_2$ 4.6–5.6 kPa (35–42 mmHg)
HCO_2 22–28 mmol l^{-1}

Standard bicarbonate

The standard bicarbonate is that value of HCO_3 measured in the plasma of fully oxygenated whole blood at a temperature of 37 °C and $PaCO_2$ 5.3 kPa (40 mmHg). It is therefore effectively an evaluation of the metabolic status only.

Base excess

This is the base concentration of whole blood measured by titration against a strong acid to pH 7.40 at a PCO_2 of 5.3 kPa (40 mmHg) at 37 °C. For negative values (base deficit) titration is carried out with a strong base. Base excess is measured in mmol l^{-1} and is an attempt to quantify the excess or deficit of HCO_3. The normal range is -1 to $+2$ mmol l^{-1} and it represents residual buffering capacity.

Technique for blood sampling

The sampling syringe and needle should contain just enough heparin (1000 units per ml) to fill the dead space. Excess or more concentrated heparin reduces the measured pH by virtue of its own acidity. Blood may be sampled by direct arterial puncture from the radial, brachial or femoral artery or withdrawn from an indwelling arterial cannula. Care should be taken to avoid air bubbles which will lead to inaccuracy of results due to gas exchange. In particular, PCO_2 will fall.

Measurement should be undertaken immediately since metabolism will continue within blood cells at room temperature thus increasing $PaCO_2$. If this is impossible the sample should be stored on ice. Diffusion of CO_2 across the walls of plastic syringes does not produce a change in CO_2 tension within the first 3 hours of storage.

Loss of water and CO_2 from stored blood samples can lead to overestimation of electrolytes and underestimation of total CO_2. Certain autoanalysers which use microsamples do not adequately protect the samples from air so that an appreciable error may occur in calculation of anion gap.

pH measurement

H ion concentration is measured by a glass electrode with a membrane which behaves as though it were permeable only to H ions and the derived pH is displayed on a meter. This is essentially an electric battery; a glass membrane with a platinum electrode dips into an electrolyte buffer and the cell is completed with a reference electrode. Surface coating of the glass with protein reduces the electrode response which may be restored by enzyme cleaning. The reference electrode is usually calomel and saturated KCl or isotonic saline forms the filling solution. The potential difference between the two electrodes is measured by a voltmeter. For good results the system needs calibrating at a pH close to that of the sample to be measured.

PCO_2 is measured by the Severinghaus electrode which incorporates a modification of this technique when a CO_2 permeable membrane is used in which the pH sensitive glass is separated from the sample by an outer membrane permeable only to CO_2.

Between this and the glass is a thin layer of HCO_3 solution. Diffusion of CO_2 and combination with water to produce H ions alters potential difference and this voltage is picked up by the reference electrode which is usually silver/silver chloride. The whole assembly is housed in an insulated jacket to preserve temperature at 37 °C.

Bicarbonate

This may be calculated by extrapolation from the pH and CO_2 using a Siggaard-Andersen (SA) nomogram.

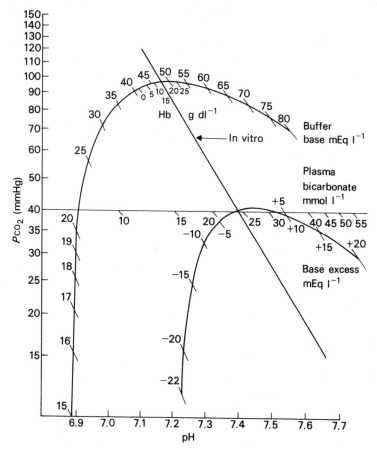

Fig. 5.3 pH, $PaCO_2$ plot on Siggaard-Andersen nomogram. T = 37 °C.

This is effectively a graphic representation of the Henderson-Hasselbalch equation, in which pH is plotted against log $PaCO_2$. Lines can be plotted on this nomogram showing the changes in pH which occur when a sample with a normal haemoglobin concentration is equilibrated with various concentrations of CO_2.

An arterial blood sample is taken and equilibrated with two gas mixtures containing different but known concentrations of CO_2 and the pH of the sample is measured. Two points are therefore plotted on the nomogram and joined by a straight line. This is called the buffer line and describes the relationship between pH and $PaCO_2$ in that particular blood sample. The pH of the patients blood is measured anaerobically in the sample. Using this pH value and the buffer line the CO_2 of the sample can be interpolated from the nomogram.

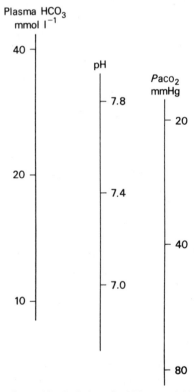

Fig. 5.4 Nomogram for rapid calculation of acid base variables.

The buffer line will cross the horizontal plasma bicarbonate line at a $PaCO_2$ of 5.3 kPa (40 mmHg), where the bicarbonate value can be read. This is the standard bicarbonate.

The metabolic component of an acid base disorder may therefore be described by the standard bicarbonate which in effect is independent of the respiratory component. However, base excess is a better parameter as it permits quantitative calculation of the dose of HCO_3 required for correction of the disturbance.

If a primary disturbance occurs in the respiratory or metabolic mechanism for H ion excretion a compensatory change takes place in the other component to restore the pH towards normal but this never is complete and metabolic compensation takes about a week to become fully effective.

Some simplified nomograms based on the Henderson-Hasselbalch equation (HHE) permit calculation of a third variable when any two have already been measured (Fig. 5.4). However, there are pitfalls to this approach in addition to those of experimental error and the need for quality control, the most important being that the HHE is only really true for very dilute aqueous solutions. Most body fluids, however, are complex molecular mixtures where water is to some extent 'compartmentalized'. Not infrequently pK calculated from the Henderson Hasselbalch equation, where all three variables have been measured is inconsistent. This is most likely to be a problem in the presence of large ionized macromolecules of proteins which bind CO_2. This phenomenon may be of practical importance in acutely ill patients if the calculation of HCO_3 uses a constant pK of 6.1. Thus the Henderson Hasselbalch equation is valid for calculation of HCO_3 in most clinical circumstances but in the case of the acutely ill, HCO_3 and total CO_2 cannot be calculated from pH and pCO_2 measurements with adequate accuracy. In these patients especially where their condition is changing rapidly CO_2 content should be measured directly to obtain HCO_3. Many hospital autoanalysers now measure total CO_2 content routinely.

Transcutaneous pCO_2

CO_2 diffuses readily through body tissues and recently developed electrodes make it possible to monitor pCO_2 continuously, non-invasively from the surface of the body. The CO_2 sensor is based on the Severinghaus electrode. The skin must be heated under the surface of the electrode to produce capillary vasodilatation and arterialisation of the blood within. A correction factor is applied to allow for the increase in pCO_2 due to the production of heat and to estimate $PaCO_2$. As this can provide continuous monitoring it may be particularly useful for predicting the onset of respiratory failure in patients not receiving ventilatory support.

Appendix to Part I

It is a capital mistake to theorize
before one has data

Memoirs of Sherlock Holmes
Sir Arthur Conan Doyle

It is now extremely easy to ask for a routine urea and electrolyte (U and E) measurement. Laboratories can process large numbers of blood samples in an autoanalyser, such as SMAC. Unfortunately, vast numbers of normal blood samples are processed for each abnormal one and there is little justification for requesting such an investigation (U and E) unless an abnormality is suspected or extremely likely, although discriminant analysis of laboratory data has been used to classify patients according to likely outcome. In any event, it is essential to know when an abnormality occurs in a routine U and E measurement, what is the total clinical picture affecting the patient at that time, and events interpreted in the light of knowledge of normal control and physiology of blood fluids. This is particularly true when a measurement of urine U and E is requested. The limitations of a random concentration of electrolytes in urine have already been pointed out. It is important in looking at the urinary excretion of electrolytes to collect the total excretion over 24 hours. In certain circumstances these can be extremely helpful in patient management, provided the clinician is selective in his requests. Biochemical testing on urine aimed to answer a specific question can be useful and cost effective. Unselected testing is a waste of resources and often gives misleading results that are ignored or misinterpreted.

Most pathologists take the view that laboratory testing is more reliable than that achieved nearer to the patient because quality control should improve analytical performance. In patients who are not receiving diuretics the guidelines for urinalysis results in Table A may be helpful. Other aspects of these abnormalities will be further discussed in the chapters in Part II.

Table A. Interpretation of urinary electrolytes.

Volume depletion	Na 0–10 mmol l^{-1}	Extrarenal salt loss
	Na > 10 mmol l^{-1}	Renal salt wasting' or adrenal insufficiency
Acute oliguria	Na 0–10 mmol l^{-1}	Pre-renal impaired perfusion
Hyponatraemia	Na 0–10 mmol l^{-1}	Severe volume depletion Oedematous states
	Na > dietary intake	Inappropriate ADH secretion Adrenal insufficiency
Hypokalaemia	K 0–10 mmol l^{-1}	Extrarenal K loss
	K > 10 mmol l^{-1}	Renal K loss
Metabolic alkalosis	Cl 0–10 mmol l^{-1}	Cl responsive alkalosis
	Cl >/= dietary intake	Cl resistant alkalosis

It is rare to request a plasma level of one particular electrolyte and more often interpretation of the plasma electrolyte profile is required. Instruments for measurement of electrolytes are now extremely sophisticated. Ion selective electrodes are capable of measuring sodium and potassium on as little as 20 microlitres of whole blood. Despite this sophistication, a result is only of value, as has already been pointed out, if the sample has been taken meticulously to avoid spurious abnormalities in results. Ion selective electrodes can also be incorporated into catheters for continuous monitoring of electrolytes such as K and Ca in blood or ISF.

There is increasing interest in new ion selective electrode machines which can be used by clinicians close to the patient's bedside for rapid estimation of Na and K. The machines are semiautomated and require calibration regularly. Biochemists are anxious that absence of quality control and skilled technicians will produce unreliable results.

In addition advances in measurement have led to estimation of the quantities of loss in the various compartments. Excluding acute blood loss, it is now possible to estimate the category and volume of fluid loss in acutely depleted patients by measurement of the haematocrit and plasma protein concentrations in venous plasma samples. In surgical patients acute depletion of body fluids occurs because plasma or extracellular fluid or a mixture of these fluids is lost. An acute loss of plasma will cause an increase in the concentration of red cells in the vascular compartment, that is to say an increase in haematocrit. In extracellular fluid loss there is no loss of plasma protein from the plasma compartment but only water and electrolyte, therefore the plasma protein concentration will rise. However, the normal range of plasma proteins is wide but measurement of both haematocrit and plasma protein concentration in a situation of acute fluid depletion can help to distinguish between a shrinkage in the plasma compartment alone and a shrinkage in the extracellular space so that logical fluid repletion can take place. The problems will be discussed more fully in Part II.

Such is the sophistication of modern methods that serial total body estimates of the absolute amounts of fat, protein, minerals and water within the body, have been developed for research purposes. After the patient has been weighed and skin fold thickness measured, total body contents of potassium, nitrogen, sodium, chloride, phosphorus can be measured with a whole body radiation counter after irradiation with fast neutrons. Calculations permit absolute values for the amounts of body fat, proteins, minerals and water. Serial measurements permit determination of the tissue composition of weight loss or gain, for example whether this is fat, muscle or water, and permit an evaluation of the effects of parenteral nutrition. This, understandably, is a method totally unsuited to routine monitoring, but it does much to advance our understanding of acute changes in critically ill patients.

Dynamic substrate metabolism can be studied by indirect calorimetry, substrate loading, measurement of AV differences using the Fick principle and isotope infusion. Increased availability of stable isotopes (deuterium, ^{13}C, ^{15}N) permits measurement of several substances at once. Isotope infusion techniques can be combined with indirect calorimetry to measure oxygen consumption and CO_2 production for calculation of RQ and are leading to better understanding of the metabolism of ill patients and more logical use of nutritional support. Before abnormalities of body water and electrolytes are discussed, there follows a number of tables indicating normal values within the human body.

TABLES

Table A.1 Composition of whole adult body.

Body weight	70 kg
Water	42 l
Total N_2	2 kg
Protein	12 kg
Sodium	5150 mmol
Potassium	3500 mmol
Chloride	2940 mmol
Calcium	32 500 mmol
Phosphate	22 600 mmol
Magnesium	1125 mmol
Fat	10.5 kg
Carbohydrate	0.5 kg

Table A.2 Normal values. There is some variation between laboratories.

Sodium	133–145	mmol l^{-1}
Potassium	3.5–4.8	mmol l^{-1}
Chloride	95–105	mmol l^{-1}
Bicarbonate	24–32	mmol l^{-1}
Urea	2.5–6.6	mmol l^{-1}
Glucose (random)	2.9–9.4	mmol l^{-1}
Creatinine	44–124	μmol l^{-1}
Uric acid	0.1–0.4	mmol l^{-1}
Albumin	35–50	g l^{-1}
Protein	60–80	g l^{-1}
Calcium	2.2–2.6	mmol l^{-1}
Magnesium	0.7–0.9	mmol l^{-1}
Phosphate	0.8–1.35	mmol l^{-1}
Creatinine clearance	40–75	ml $min^{-1} m^{-2}$
Arterial blood gases		
pH	7.36–7.42	
PCO_2	4.4–6.2	kPa
Base excess	+/− 2.3	mmol l^{-1}
Standard bicarbonate	23–28	mmol l^{-1}
PO_2	12–15	kPa
Hydrogen ion	36–44	mmol l^{-1}
Anions	12+/− 4	mEq l^{-1}
Osmolality	280–295	mosmol kg^{-1}
Colloid osmotic pressure	20–25	mmHg

Table A.3 Intracellular concentrations.

Sodium	10–35 mmol l^{-1}
Potassium	150 mmol l^{-1}
Magnesium	12–18 mmol l^{-1}
Bicarbonate	10–20 mmol l^{-1}
Chloride	5 mmol l^{-1}
Phosphate	31 mmol l^{-1}

Table A.4 Composition of secretions. If visible sweating occurs there is a marked increase in its sodium and chloride concentration.

Secretion	Na	K	Cl	HCO$_3$	Volume
		mmol l^{-1}			(litres daily)
Parotid saliva	112	19	40	–	1.5
Gastric juice	50	15	140	0–15	2–3
Pancreatic juice	130	5	55	110	0.5–1
Bile	145	5	100	38	0.5–1
Ileal juice	140	11	70	Variable	–
Normal stool	20–40	30–60	20	–	0.1
Diarrhoea	30–140	30–70	–	20–80	Variable
Insensible sweat	12	10	12	–	0.5

Table A.5 Normal urine values.

Specific gravity	1003–1030
Creatinine/24 hours	8.8–17.7 mmol
Urea clearance	60–95 ml min^{-1}
Glomerular filtration rate (GFR)	105–140 ml min^{-1}
Osmolality	30–1400 mosmol kg^{-1}

Table A.6 Principal urine constituents. In normals these are greatly influenced by dietary intake. All these values will be modified depending upon renal function, the degree of catabolism and the presence of infection.

	Normal	Effect of operation
Water ml	1500	500
	(1000–2500)	
Sodium mmol	70–160	5–20
Potassium mmol	40–120	90–180
Urea g	16–350	20–60

Table A.7 Typical nitrogen composition of normal urine on 90 g protein diet.

	g 24 hour^{-1}
Total	14.40
Urea	12.50
Creatinine	0.67
Ammonia	0.44
Uric acid	0.23
Other	0.56

APPENDIX TO PART 1

Table A.8 Sodium potassium, N_2 and energy content of intravenous fluids.

Solution	Na mmol l^{-1}	K mmol l^{-1}	N_2 g l^{-1}	Energy kcal l^{-1}	Energy MJ l^{-1}
Whole blood (fresh)	77	3	6	–	–
Packed cells (PCV 60%)	55	2	4	–	–
Human Albumin Fraction 5%	145	2	8	–	–
Human Albumin Salt poor 20%	130	10	32	–	–
Haemaccel 3.5%	145	5.1	6.3	–	–
Gelofusine 4% gelatin	154	0.4	–	–	–
HPPF	150	2	6	–	–
HAS (45g l^{-1} protein)	153	<2			
Dextran 70 in 0.9% NaCl	150	–	–	–	–
Dextran 70 in 5% Dext.	–	–	–	190	0.79
Polyfusor phosphates	162	19	–	–	–
Dextrose 10%	–	–	–	380	1.59
Sodium chloride 1.8%	308	–	–	–	–
Viaflex NaCl 0.45%	75	–	–	–	–
Baxter Travenol NaCl 0.225%	37.5	–	–	–	–
NaHCO$_3$ 1.4%	167	–	–	–	–
Polyfusor NaHCO$_3$ 4.2%	500	–	–	–	–
Polyfusor NaHCO$_3$ 8.4%	1000	–	–	–	–
Viaflex Hartmanns	131	5	–	–	–
Viaflex 0.18% NaCl, 4% dextrose 0.2% KCl	30	27	–	150	0.63
Viaflex 0.18% NaCl, 4% dextrose	30	–	–	150	0.63
Viaflex 0.9% NaCl + 0.2% KCl	150	27	–	–	–
Viaflex 0.9% NaCl	150	–	–	–	–
Viaflex 5% dextrose + KCl 0.2%	–	27	–	190	0.79
Viaflex 5% dextrose	–	–	–	190	0.79
Aminoplex 12	35	30	12	–	–
Aminoplex 14	35	30	14	–	–
Intralipid 10%	–	–	–	1000	4.62
Intralipid 20%	–	–	–	2000	8.4
Dextrose 20%	–	–	–	750	3.14
Dextrose 40%	–	–	–	1500	6.28
Glucoplex 1600	50	30	–	1600	6.72
Glucoplex 1000	50	30	–	1000	4.2
Vamin glucose	50	20	9.4	650	2.75
Vamin N	50	20	9.4	236	1.0
Potassium phosphate 1 mmol HPO$_4$, 2 mmol K ml^{-1}	–	2000	–	–	–
Dopram solution 2 mg ml^{-1} in 5% dextrose	–	–	–	190	0.97
Metronidazole	130	–	–	–	–

PART II

Chapter 6
Abnormal Water Balance, and Disturbances of Osmolality

Water balance is closely related to Na balance and to separate the two is to some extent a false distinction. Water balance is largely controlled by Na balance. Water deprivation produces a hypertonic state and water intoxication a hypotonic state. Usually these are synonymous with hyper and hypotonic syndromes. Strictly speaking dehydration means loss of water only. Unfortunately sodium depletion always causes reduction in ECF volume and this is often referred to as dehydration. The distinction between the two is illustrated in Table 6.1 taken from Leaf (1984).

Table 6.1 Comparison of water lack and salt deficiency.

Feature	Water lack	Salt deficiency
Clinical		
Thirst	Intense	Unremarkable
Skin turgor	Normal	Reduced
Pulse	Normal	Rapid
Blood pressure	Normal	Low
Orthostatic hypotension	Absent	Present
Laboratory		
Urine volume	Small	Unremarkable
Urine concentration	Maximal	Unremarkable
Haemoconcentration	Slight	Present
Blood urea	High normal	Increased
Plasma sodium & osmolality	Increased	Reduced

WATER DEPLETION

This gives rise to thirst so that pure water depletion is often associated with impaired consciousness or occurs in those who cannot voluntarily increase their water intake such as the very young and the very old. A patient willing and able to increase water intake can avoid hypertonicity in the complete absence of antidiuretic hormone (ADH) or the ability to concentrate his urine.

Thirst

Thirst is basic to our existence. It acts as an emergency mechanism to repair acute fluid deficits. Normally drinking is anticipatory of future needs and depends on diet, habit and innate circadian rhythms. It is usually dependent on the current need for water. Symptomatic thirst occurs in response to fluid loss from ECF or ICF. These two fluid compartments have their own controlling mechanisms which act independently of each other but have an additive effect on thirst if activated together.

The immediate stimulus for cellular control is a reduction in the volume of the cells in particular those in the hypothalamus. Cellular volume probably fluctuates between an upper satiety volume and a lower threshold level at which thirst occurs. Normally there is a reserve of body fluid so that the level of hydration is above that at which thirst is stimulated.

Hypovolaemia is the second major stimulus to thirst production and aids in maintenance of plasma volume. It is in ECF or hypovolaemic thirst that the renin-angiotensin system participates to a variable extent, although a nephrectomized animal can still respond to some ECF stimuli. Involvement of the renin-angiotensin system in hypovolaemic thirst enables the subject to respond to ECF dehydration in two ways.

1 It promotes drinking.

2 It produces Na and water retention by an angiotensin stimulated increase in aldosterone production. This will also increase Na appetite.

Symptomatic thirst therefore occurs in response to fluid loss. However, thirst may also be pathological in which case it persists although the body is normally or over-hydrated. For confirmation of this diagnosis not only must the fluid intake be shown to be disproportionately high but the trend in fluid balance should be measured. Repeated measurement of weight and urine volume should be made and the presence of oedema sought. Pathological thirst occurs in some cases of diabetes insipidus, in compulsive water drinking, stimulation of the neural thirst mechanism by a high plasma renin, in hypercalcaemia and hypokalaemia, in congestive cardiac failure and in manic depressives treated with lithium. In some cases of chronic renal failure with high plasma renin, angiotensin II and aldosterone there may be marked thirst which may be due either to a direct action of renin on the thirst centre or due to reduction in blood volume. Thirst is then often abolished by bilateral nephrectomy.

Thirst occurs when 2% of the body weight has been lost as water (1.5 l in a 70 kg man).

When water is depleted relative to electrolytes the tonicity of body fluids is increased resulting in stimulation of ADH from the posterior pituitary in addition to thirst. Increased ADH increases reabsorption of free water in the distal renal tubule. A rise of 1 mosmol kg^{-1} in plasma osmolality increases urine osmolality by nearly 100 mosmol. This has been discussed more fully in Chapter 1. Synapses involved in ADH release are cholinergic and thus stimulated by nicotine and acetylcholine. Water depletion will only occur when the thirst mechanism fails.

It has recently been shown that in healthy elderly men subjected to water deprivation there is greater increase in plasma osmolality, sodium and vasopressin levels but less *thirst* and lower urine osmolality compared to younger men. There is a deficit in thirst and water intake although vasopressin and osmoreceptor responsiveness is maintained or even increased. This probably reflects a renal cause for reduced concentrating ability. The younger controls were thirstier, drank more and corrected their water deficit in a shorter period of time. Nevertheless basal plasma sodium and osmolality in the elderly were normal *but* mild stresses such as diarrhoea or infection could lead to rapid dehydration with mental confusion and further reduction in water intake. Thus the sensation of thirst is decreased, osmoregulated vasopressin secretion is increased and renal response to vasopressin blunted. The mechanism may be diffuse cerebrovascular disease or reduction in sensitivity of the osmoreceptor involved in thirst appreciation. Against this is the age related increase in sensitivity of the osmoreceptor regulating vasopressin secretion. The possibility remains that there are two osmoreceptors, one regulating thirst and the other vasopressin secretion. This is supported by animal work.

The elderly may equally have a deficient response to ECF volume depletion. There is certainly reduced baroreceptor sensitivity and regulation of ADH release in the elderly but severe hypovolaemia is required to initiate thirst in man. The most likely explanation for this phenomenon, therefore, is reduced urine concentrating ability and renal responsiveness to vasopressin in the elderly.

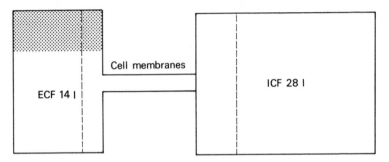

Fig. 6.1 Water deficiency. The effect of 10 litres water loss on body fluid compartments. The dotted lines show the effect of the water loss equally spread between the compartments resulting in an equal increase in osmolality in ECF and ICF. The shaded area denotes the intravascular compartment.

As water is freely permeable across cell membranes it is lost from all compartments and plasma volume is initially maintained (Fig. 6.1). This contrasts sharply with the situation in which a combined loss of salt and water occurs. Peripheral circulatory failure in pure water depletion is a very late manifestation. In fact the plasma compartment may not share to the same extent in the fluid loss because of the increase in colloid osmotic pressure which occurs as water depletion continues. This holds water within the intravascular compartment so that only about 12% of the total water loss is incurred by this compartment.

Biochemical abnormalities

1 Raised Na concentration. The commonest cause of a raised plasma Na concentration is severe water depletion with ECF shrinkage. Values greater than 150 mmol l^{-1} are virtually diagnostic of water loss provided that:
(a) The patient is not receiving an infusion of hypertonic saline.
(b) The patient is not undergoing dialysis against a fluid of abnormally high Na concentration.

Plasma Na begins to rise in an adult when 3 or more litres of water have been lost.

2 Plasma chloride and urea concentrations increase.

3 Plasma protein levels and haematocrit rise.

4 Urine output falls (less than 500 ml daily) and this urine will have a high specific gravity, osmolality and high urea concentration.

Clinical signs and symptoms of water depletion

Thirst.

Sunken cheeks and eyes.

Weakness.

Dryness of mucous membranes.

Reduction of salivary and bronchial secretions.

Loss of skin turgor.

Weight loss.

Mental disturbances occur as total body water (TBW) deficit approaches 10%. Many of the effects of hyperosmolality in water deficiency are due to cerebral dehydration when acute osmolar changes are more important than chronic changes since in the latter situation the brain is able to generate osmotically active substances (idiogenic osmoles) and hence reduce cellular dehydration. Cerebral dehydration may tear vessels especially in small infants given mannitol and hence may produce subdural or intracerebral haemorrhage. Abnormal neurotransmitter release, reduced energy production or brain cell volume changes depress conscious level and increasing hypertonicity leads to irritability, twitching and convulsions. In infants with 5% or more fluid loss due to gastroenteritis there is often no associated disturbance of osmolality although 25% develop hypo-osmolar dehydration implying sodium loss as well.

Causes of pure water loss

Decreased intake

Infancy.

Old age.

Unconsciousness.

Dysphagia — carcinoma of the oesophagus.

Severe nausea and vomiting.

'Nil by mouth' regime of perioperative surgical patients with inadequate parenteral fluids. Postoperatively if intravenous fluid is withheld a patient will drink about 25 ml kg^{-1} water daily. This amount should therefore be supplied.

Hyperosmotic diets — these result in a relative water deficit.

Pathological hypodipsia due to lesions in the supraoptico-hypophyseal system are extremely rare but may be associated with disturbance of temperature, appetite and sleep. Insensitivity to pain may accompany this with a possibility of good response to naloxone. When ADH secretion is also affected hyperosmolality may be extreme.

Opiates appear to be concerned with ADH secretion. If volunteers are deprived of water, infusion of an analogue of met-enkephalin (an endogenous opiate peptide) produces a diuresis attenuated by naloxone. In hydrated subjects this does not occur. Following administration of synthetic met-enkephalin, plasma immunoreactive ADH levels fail to increase despite hypertonic saline infusion. Therefore opiates appear to suppress osmotically mediated release of ADH. However, beta-endorphin can cause secretion of ADH in man: an effect which can be reversed by naloxone. The overall effect of opiate peptides therefore requires further elucidation.

Increased water loss

LUNGS

Marked hyperventilation may accompany fever and cerebral damage after head injury thus increasing water loss from the lung. In addition extreme breathlessness may limit intake. Patients receiving intermittent positive pressure ventilation must be adequately humidified otherwise considerable water loss will occur.

SKIN

Evaporation of water from the skin surface is a normal temperature regulating mechanism. However, in high environmental temperatures or thyrotoxicosis several litres of hypotonic fluid may be lost daily through the skin.

RENAL LOSSES

Increased urine flow
1 Osmotic diuresis due to mannitol, hyperglycaemia or urea. This has been discussed in Chapter 1. The presence of a non reabsorbable solute in the glomerular filtrate results in passage of a large volume of hypotonic urine with subsequent net water loss.
2 Diuretics. These are discussed in Chapter 14.
Renal tubular defects. These are discussed in Chapter 11.
Diabetes insipidus (DI). This disorder is due either to a defect in ADH production or a defect in the renal response to ADH (nephrogenic DI).

Central diabetes insipidus

This is due to damage in the region of the hypothalamic pituitary system.

CAUSES OF CENTRAL DIABETES INSIPIDUS

Tumour:
  primary — usually a craniopharyngioma
  secondary.
Trauma:
  head injury
  post hypophysectomy.
Infection:
  meningitis
  encephalitis.
Others:
  cerebral sarcoidosis
  Hand-Schüller-Christian disease.
Some of these are accompanied by a disturbance of thirst. Thirst is pathological and produces marked polydipsia and this may be accompanied by malaise and nausea.

Traumatic DI accounts for about 40% of all cases of DI and has a triphasic onset.
1 Polyuria occurs due to insufficient ADH.
2 Uncontrolled release of ADH may then occur at a time when the high fluid intake prescribed for the polyuric phase could result in hyponatraemia and water intoxication.

3 1–7 days later DI reappears. This may be permanent or transient since regeneration of ADH production can occur months after the initial injury.

DRUGS

Certain drugs such as alcohol, lithium and narcotic analgesics inhibit ADH release.

In central diabetes insipidus the pathology may also produce a disorder of thirst and this will determine the clinical picture. Patients with reduced or absent ADH secretion and normal thirst mechanisms present with polyuria, polydipsia and normal plasma osmolality. Patients with reduced or absent ADH and impaired thirst mechanisms do not have polydipsia but develop plasma hyperosmolality.

Usually a patient with DI can concentrate his urine to a maximum of 300 mosmol kg^{-1} but will have some polyuria and excess water loss. Most patients excrete a dilute urine despite dehydration but can concentrate their urine in response to exogenous ADH. Nocturia usually occurs and helps to distinguish DI from psychogenic polydipsia.

DIAGNOSIS OF CENTRAL DIABETES INSIPIDUS

1 The time-hallowed water deprivation test may reveal the diagnosis. Fluid restriction is continued until the patient has lost 3–5% of his body weight or urine osmolality is constant in 3 hourly urine specimens. Five units of subcutaneous ADH are then given and urine is collected for 2 hours. In DI this maneouvre will produce an increase in urine osmolality. In nephrogenic DI or psychogenic polydipsia there is no change in urine osmolality.
2 Water deprivation is not acceptable to many clinicians who use the response to intramuscular vasopressin tannate in oil.
3 This has now to a large extent been replaced by intranasal synthetic deamino-D-arginine-vasopressin (desmopressin, DDAVP). This form of ADH is also a valuable agent for therapy.
4 Infusion of hypertonic saline can be used to osmotically stimulate ADH release and separate those patients with normal osmo-

regulation of ADH secretion from those with central DI. If causes of polyuria cannot be established by simple dehydration then measurement of ADH during infusion of hypertonic saline will distinguish between those with normal osmotically stimulated ADH release and those without. In normal individuals a strong positive correlation exists between plasma osmolality and plasma ADH levels whereas a subnormal ADH response occurs in central DI.

Nephrogenic diabetes insipidus

Two mechanisms may be involved:
1 The distal tubule and collecting duct lose the capacity to respond to ADH although the generation of the medullary osmotic gradient is normal.
2 Generation of the osmotic gradient itself may be impaired. Some drugs may reduce renal concentrating ability (Table 6.2).
The vasopressin defect in nephrogenic DI is not confined to the kidney but is shown in other tissues such as vascular endothelium and hepatic sinusoids and can be demonstrated by decreased factor VIII coagulant activity and factor VIII related antigen. This latter may help to identify families at risk.

Table 6.2 Drugs which reduce renal concentrating ability.

Lithium
Tetracyclines
Methoxyflurane (fluoride ions)
Sulphonylureas
Amphotericin B
Colchicine

A case of nephrogenic DI has been reported due to the combination of triamterene and hydrochlorothiazide.

In addition hypercalcaemia and hypokalaemia reduce concentrating ability. Transient vasopressin resistant DI can occur in late pregnancy. Management of DI is considered after general treatment of water deficiency.

CALCULATION OF WATER DEFICIT

Body water deficit = normal body water — current body water

Normal body water can be calculated from a knowledge of approximate normal body weight:

Normal body water = 0.6 × normal body weight (kg)

Current body water may be calculated thus:

$$\frac{\text{Normal plasma Na} \times \text{normal body water}}{\text{measured plasma Na}}$$

For example in a 70 kg man with a normal plasma Na (140 mmol l^{-1}) and a measured Na of 150 mmol l^{-1}

$$\text{Body water deficit} = (0.6 \times 70) - \frac{(140 \times 0.6 \times 70)}{150}$$

$$= 42 - \frac{(140 \times 42)}{50}$$

$$= 42 - 39.2 = 2.8 \text{ litres}$$

Since the deficit is one of pure water total body solute remains the same. Hence:

Normal body water × normal osmolality
= present body water × present osmolality.

Normal body water can be calculated as discussed and normal osmolality is 285 mosmol kg^{-1}.

If for example a 70 kg man has a measured osmolality of 310 mosmol kg^{-1} this gives the following:

$$0.6 \times 70 \times 285 = \text{present body water} \times 310$$

$$\text{Present body water} = \frac{0.6 \times 70 \times 285}{310}$$

$$= \frac{1197}{31} = 38.6 \text{ litres}$$

Therefore:

$$\text{Body water deficit} = (0.6 \times 70) - 38.6$$

$$= 42 - 38.6 = 3.4 \text{ litres.}$$

A simplified version of this depends on the principle that the percent reduction in body water will be proportional to the percent increase in plasma Na. However, in some cases of dehydration plasma Na is normal and this is unhelpful.

Treatment

It is essential to distinguish clinically between water deficit and Na excess (see also Chapter 7). Clearly treatment must be that of the underlying condition in addition to replacing deficits and continuing losses.

Water replacement

Water may be given orally if the patient can drink or via a nasogastric tube if it can be absorbed and there is no risk to the airway. 48 hours should be allowed for correcting a water deficit. It is very important to avoid overhydration since the cerebral idiogenic osmoles which were produced during dehydration will now act to draw water into cells and produce cerebral oedema with convulsions and a deteriorating conscious level. More often this occurs with overenthusiastic intravenous therapy since this route of repletion is almost always needed in more severe cases.

Solutions available include hypotonic saline such as 0.45% NaCl and 5% dextrose. Prior to infusion 5% dextrose is isotonic: following intravenous infusion the dextrose is metabolized to carbon dioxide (CO_2) and water so that water is effectively supplied without electrolytes. A simple safe 'push technique' has been used in moribund children with gastroenteritis. 20–50 ml/kg 5% dextrose in 0.9% saline is given into the femoral vein over 15–20 minutes in 30 ml boluses. This may lead to sufficient improvement to continue with an oral rehydration solution.

Oral rehydration therapy is a very important, powerful tool to reduce mortality from diarrhoeal illness especially in underdeveloped countries. The exact content of the solution remains to be finalized where nutritional depletion is also rife but WHO recommendations for oral rehydration solutions (ORS) are as follows:

mmol l^{-1} water
   Sodium 90
   Potassium 20

Chloride 80
Bicarbonate 30
Glucose 111

ORS should be refridgerated if possible after making up as bicarbonate stability may decrease with increased temperature. Some of the commercially available ORS are flavoured with citric acid which causes a sharp reduction in bicarbonate concentration as it is converted to CO_2 and water. The value of HCO_3 is anyway questioned. Addition of honey to these solutions was found to reduce the duration of bacterial diarrhoea without increasing the duration of non-bacterial diarrhoea.

Severe dehydration during episodes of diarrhoea as well as having a high mortality also increases the incidence of cataract formation in India. Mothers can prepare a rice salt ORS in the home in Bangladesh. In a study of 1330 children of less than 3 years of age 21% had hyponatraemic fluid loss, 73% isonatraemic loss and 6.4% were hypernatraemic and this increased with age. The incidence of hyponatraemia was directly related to nutritional state and also correlated with higher fatality.

Use of the UNICEF/WHO ORS, originally designed for treatment of cholera has been questioned for treatment of gastroenteritis in the Western world. Therefore BPC fluid; Na 35 mmol, K 20 mmol, HCO_3 18 mmol and dextrose 200 may be more suitable. Details are beyond the scope of this book. Drug treatment with doperamide with its opiate action on the bowel may reduce the duration of diarrhoea, and chlorpromazine has been valuable in cholera (p. 311).

MONITORING

The patient should be weighed daily and a fluid balance accurately measured. In an adult the aim is a urine volume of at least 1000 ml. Unfortunately, in the absence of catheterization accurate urine measurement may be surprisingly difficult.

The conscious level is a valuable guide to the rate and adequacy of replacement.

Laboratory tests. Repeated measurements should be made of haematocrit, urea and electrolyte levels to ensure gradual controlled correction.

Following calculation of the fluid deficit, replacement should be as follows:

0.33% in the first 6 hours.

0.33% in the subsequent 18 hours.

0.33% in the next 24 hours.

In addition to the deficit, allowance must also be made for continuing losses and basal requirements. However, if the patient is elderly an additional 24–48 hours may be required to safely correct the deficit. If the conscious level first improves and then worsens again cerebral oedema should be suspected which if severe may require therapy with intravenous mannitol to reduce intracranial pressure.

Specific treatment: central diabetes insipidus

1 Antidiuretic hormone.
2 ADH analogues.
3 Other drugs.

In the acute situation sterile aqueous vasopressin (ADH, pitressin) 5–20 units intramuscularly should be given at least twice daily. This solution in the early days was contaminated by alpha melanostimulating hormone and had significant ACTH releasing activity. Pitressin may be dangerous in patients with heart disease because it produces arteriolar constriction with hypertension and acute venous (splanchnic) constriction which increases circulating blood volume. Urine output and thirst are helpful in timing of dosage. This drug is also used for bleeding oesophageal varices to reduce portal venous pressure secondary to splanchnic vasoconstriction. Continuous ECG recording during infusion should be used to detect myocardial ischaemia. Colicky abdominal pain, pallor and bowel evacuation occur with hyponatraemia and oliguria. Propranolol will also reduce portal pressure by inhibition of beta 2 receptors on the hepatic artery and mesenteric arterioles (thus decreasing splanchnic blood flow).

ADH ANALOGUES

1-deamin 8-D-arginine vasopressin (DDAVP) is available as a nasal solution. It has twice the duration of action of lysine vasopressin and may be given in a dose of 10–20 µg once or twice daily.

DDAVP is supplied in a dropper bottle with a calibrated catheter for intranasal instillation. Each 0.1 ml contains 10 µg and detailed instructions are supplied for the method of administration. Although the oxytoxic effect of this solution is very low it should be used with great caution in pregnant patients.

DDAVP injection is also available, and may be given intramuscularly or intravenously in a dose of 1–4 µg daily. DDAVP raises plasma concentration of factor VIII and has a role in treatment of haemophilia and von Willebrands disease. It also shortens bleeding time in uraemia for about 6 hours and may prevent surgical bleeding in uraemic patients with prolonged bleeding times. DDAVP does not cross the blood brain barrier when given i.v.

Lysine vasopressin is shorter acting than DDAVP and is given intranasally 5–20 units, 3–7 times daily. It is available in a metered dose spray (2.5 units per squeeze).

Glypressin (terlipressin) is triglycyl-lysine-vasopressin. Intravenous injection results in slow cleavage of the glycyl residues by enzymic action, releasing vasopressin. Glypressin therefore provides a circulating depot which releases vasopressin at a constant rate. The drug is claimed to be more effective than vasopressin with fewer complications although the dangers of generalized vasoconstriction cannot be overlooked and it requires further evaluation.

Some drugs can decrease renal diluting capacity and may be useful in mild cases of central DI.

1 Thiazide diuretics (useful for nephrogenic DI also).
2 Chlorpropamide.
3 Carbamazepine.
4 Clofibrate.

Chronic administration of thiazide diuretics can reduce urine volume by 50%. Other diuretics or salt restriction produce a similar result. However, if salt is added to therapy this antidiuresis does not occur. In fact a negative salt balance (Na depletion) is essential for this effect. A reduction in GFR occurs with enhanced Na reabsorption in the proximal tubule so that a reduced amount of filtrate reaches the distal tubule and urine volume is reduced. There is no effect on plasma ADH level. Chlorpropamide has no effect on ADH levels either and is ineffective in nephrogenic DI. The effect of chlorpropamide is to enhance the peripheral action of ADH by sensitizing the renal tubules to the action of endogenous ADH. Chlorpropamide is given in a dose of 250 mg once to twice daily initially increasing to a maximum of 1000 mg daily. This dose also increases thirst. Chlorpropamide may be used in conjunction with a thiazide diuretic. Carbamazepine probably acts in the same way as chlorpropamide and is used in a dose of 200 mg once or twice daily. Amiloride may reduce lithium induced polyuria partly by blunting the inhibitory effect of lithium on water transport in the renal collecting duct. This is somewhat contentious and the dose requires careful monitoring.

It must be stressed that pure water depletion is rare and much more commonly mixed salt and water depletion occur.

WATER EXCESS; WATER INTOXICATION

When the body accumulates excess water compared to electrolyte, hypo-osmolality occurs. Initially this excess water is distributed throughout the body and little change in plasma electrolyte levels occurs. Progressive water accumulation leads to a small increase in ECF volume without oedema and a progressive fall in plasma Na. Very low levels of plasma Na (less than 120 mmol l^{-1}) usually imply some ECF salt loss as well as water excess.

Clinical features

Symptoms depend on the severity and speed of development of the hypo-osmolality. Plasma Na less than 120 mmol l^{-1} is associated with neurological abnormalities only partly related to the development of cerebral oedema. The following clinical features may be present:

Confusion, disorientation.
Restlessness.
Headache, cerebral oedema.
Convulsions, nocturnal jerks, muscular cramps.
Coma.

Biochemical features

1 Low plasma Na and Cl.
2 Low haematocrit, haemoglobin and plasma proteins.
3 In most instances urine Na is greater than 20 mmol l^{-1}.

Causes of water excess

Increased water intake

1 Pathological polydipsia.
2 Ill advised fluid regimes. During the postoperative period when the stress response to trauma occurs, the kidney has a reduced ability to excrete free water. Administration of 5% dextrose in this situation will effectively supply more water and worsen the situation. If blood or ECF volume deficits have been inadequately replaced then the production of ADH and aldosterone will be stimulated. Accidental oral water intoxication can occur without psychiatric disturbance; one case resulted in fatal cerebral oedema. Even mild oral overhydration may precipitate grand mal convulsions in epileptics.
3 Water absorption during bladder irrigation is rare now that 1.5% glycine solutions are used for irrigation. This solution is almost isotonic and reduces the dangers of haemolysis following water absorption. However, severe hyponatraemia and ECG changes which include a wide QRS of increased amplitude, T wave inversion and bradycardia are still reported. Perforation of the bladder during diathermy can lead to an enormous leak of fluid into the peritoneal cavity and severe water intoxication.

Inability to excrete water normally

1 Inappropriate secretion of ADH from the hypothalamus or pituitary region. Plasma osmolality is low.
2 Inappropriate secretion of a similar peptide from a neoplasm for example carcinoma of the bronchus or pancreas.
3 Adrenal insufficiency.
4 Hypothyroidism.
5 Pain and emotion.
6 Drugs.
(a) Those which act centrally to reduce renal diluting capacity:
Nicotine.
Narcotics.
Sulphonylureas.
Clofibrate.
Carbamazepine.
Vincristine.

(b) Those which act on the kidney to reduce renal diluting capacity:

Diuretics.

Sulphonylureas.

Biguanides.

Anti-inflammatory drugs such as aspirin, paracetamol.

Diazoxide.

Experimentally angiotensin II when injected into specific regions of the hypothalamus may produce either drinking of water, release of ADH or rises in blood pressure.

Oxytocic infusions can cause profound hyponatraemia and water intoxication partly due to their being made up in 5% dextrose. Even low doses of oxytocin are antidiuretic ($2–5$ mU min^{-1}) and this effect occurs within 10–15 minutes of the onset of the infusion. Overtransfusion contributes to the severity of the problem. This cause of water intoxication is becoming less likely now that prostaglandins are used more often for induction of labour. An isotonic vehicle should always be used in small volumes. Administration of pethidine may further increase ADH. Iatrogenic hyponatraemia of the newborn is strongly related to infusion of hypotonic fluids to the mother during labour.

Transfer of water from cells

Strictly this is not a situation of true water excess. It occurs in the following circumstances:

1 Uncontrolled diabetes mellitus.

2 Administration of mannitol.

A sudden rise in blood glucose draws intracellular water into ECF diluting plasma constituents. For every 3 mmol l^{-1} elevation of plasma glucose there is a decrease of 1 mmol l^{-1} in plasma Na by osmotic dilution. However, such circumstances lead to an osmotic diuresis such that dehydration and increased plasma osmolality result. Transfer of water from cells therefore is a very short lived phenomenon.

Polydipsia

Pathological polydipsia may be due to a psychiatric disturbance. Compulsive water drinking occurs in a wide range of conditions from mild neurosis to psychosis but is not uncommon in chronic schizophrenics. In the USA the incidence of compulsive water drinking in state mental hospitals is 7–18% and half of these have some complication of water intoxication. In this group of patients major motor epilepsy is the commonest presenting feature (in 80%). A very few of these may have defined inappropriate ADH secretion possibly precipitated by treatment such as chlorpromazine and amitriptyline. The neuroleptic malignant syndrome can be associated with inappropriate antidiuresis and psychogenic polydipsia. There is some dopaminergic control of ADH release and thirst behaviour such that dopamine stimulates thirst. Many psychotropic drugs affect dopamine receptors. Hypo-osmolality can occur if the renal diluting capacity is impaired by drugs or renal disease. A central nervous system lesion such as tumour, trauma or inflammation within the hypothalamic thirst area will produce the same result.

Hypercalcaemia and hypokalaemia increase thirst but the associated polyuria prevents hypo-osmolality. Drugs such as thioridazine and chlorpropamide increase thirst. Any circumstance in which the function of the countercurrent mechanism is impaired will reduce renal diluting capacity.

ADH excess syndromes

ADH secretion may occur in the presence of hypo-osmolality when nonosmotic stimuli lower the threshold of secretion. This may occur when the effective ECF volume is reduced in oedematous states for example heart failure, hypoalbuminaemia, venous and lymphatic obstruction. Intravenous administration of somatostatin may stimulate normal ADH release. In acute alcohol withdrawal brain water is increased probably due to increased vasopressin.

However, inappropriate ADH secretion is uncommon and every effort must be made to exclude other diagnoses, in particular impaired adrenal function.

Conditions associated with pathological ADH production

PULMONARY

Carcinoma.
Pneumonia.

Abscess.
Aspergillosis.
Tuberculosis.
Other intrathoracic conditions may also be associated with this syndrome: these include thymoma, postal mitral valvotomy, artificial ventilation and severe prolonged asthma and open heart surgery.

CENTRAL NERVOUS SYSTEM DISEASE

Viral and bacterial encephalitis and meningitis.
Cerebrovascular accident.
Cerebral tumour, abscess.
Trauma.
Guillain Barré syndrome.
Excess ADH production may accompany other hypothalamic disturbances such as control of temperature, appetite and pain.

DRUGS

Vincristine.
Thiazide diuretics.
Carbamazepine.
Dothiepin.
Other antidepressant drugs.

MISCELLANEOUS

Leukaemia, Hodgkin's disease.
Porphyria.
Stress.
Hypoxia.
Halothane anaesthesia.
Surgery for craniofacial anomalies.
Infectious mononucleosis.
Although many drugs reduce renal ability to produce dilute urine there is little evidence of an abnormal ADH level. Traditionally carcinoma of the bronchus was the commonest cause of this syndrome, but in a recent series chest infection was a commoner cause and all patients had a raised ADH level. There is increasing

interest in the role of the renin angiotensin system in regulation of thirst, ADH release and sodium homeostasis. It appears that a separate intrinsic cerebral renin angiotensin system exists in which angiotensin mediates ADH release.

The increase in ADH results initially in volume expansion which leads to the following:

1 Increased renal perfusion, increased GFR and decreased proximal tubular reabsorption of Na and water.

2 Increased renal medullary blood flow and washout of the hypertonic medulla.

3 Suppression of aldosterone.

4 Hyponatraemia per se interferes directly with the action of ADH on collecting ducts, with the result that a Na diuresis occurs and a new steady state is set up. Volume expansion is therefore corrected but hyponatraemia remains. This is asymptomatic hyponatraemia unless water intoxication develops. Low blood urea also occurs, partly due to haemodilution and partly due to increased excretion as a consequence of the water expanded state.

There has been some confusion regarding measured levels of ADH in these conditions, but the following are good criteria for diagnosing inappropriate ADH secretion:

1 Hyponatraemia with reduced plasma and ECF osmolality.

2 Continued Na excretion with urine osmolality inappropriately high for plasma osmolality.

3 Normal ECF volume.

4 Urine not maximally dilute.

5 Normal renal and adrenal function.

Variation in assay methods and normal ranges and failure to correlate ADH level to plasma osmolality may account for some reports of the totally erratic ADH production. In this syndrome it does seem that some response to osmotic stimuli remains and there is evidence for the resetting of the osmostat to a new steady state with normal operation of other factors which influence water excretion. At least three different types of osmoregulatory defects have been found, none of which is specific for any disease process. It is important to remember that inappropriate ADH secretion is rare and much more often hyponatraemia is due to iatrogenic water intoxication when 5% dextrose is infused in excessive amounts as the following example shows.

An 83 year old lady who had previously been on diuretics but whose plasma sodium on admission was 140 mmol l^{-1} underwent an abdomino-perineal resection. Five days later she was completely unrousable but afebrile with blood urea 4.8, Na 106, K 4.3, Cl 75, HCO_3 22 mmol l^{-1} having been in considerable positive fluid balance since operation. In view of her tendency to retain sodium most of the i.v. fluids administered had been in the form of 5% dextrose. She failed to respond to fluid restriction alone over 24 hours and was treated with frusemide, 1.8% saline and added potassium with a good response. Two days after this treatment she demanded to go home.

In adrenal insufficiency and hypothyroidism there is inability to excrete a water load which responds to specific treatment of the underlying condition.

Treatment

For all causes of water excess, treatment of the underlying cause is most important. Mild cases of water excess may be treated by withholding water and permitting only 600–1000 ml daily to cover insensible losses. Many of the causes of inappropriate ADH secretion are short lived and self limiting. If severe life threatening central nervous system symptoms develop, however, such as convulsions, simple water restriction will be inadequate. In such circumstances one approach is to give a combination of diuretic such as frusemide with a small volume of hypertonic saline (1.8 or 3% NaCl). The hypertonic saline will expand ECF volume hence the importance of the diuretic to induce a negative water balance. In an emergency 8.4% $NaHCO_3$ solution is a more readily available hypertonic solution. An alternative is haemofiltration or haemodialysis to remove water. In Australia even more aggressive therapy with 29.2% saline has been used to control convulsions in this situation. This must be administered through a central line. Any form of treatment with hypertonic saline is dangerous in the elderly because of the risk of heart failure due to circulatory overload. In addition patients are already volume expanded and in the presence of normal renal function the

Na will be excreted so that hyponatraemia and cerebral oedema may recur. Alternative forms of therapy for treatment of chronic or persistent inappropriate ADH secretion should be reserved for situations in which water restriction fails:

1 Lithium.
2 Demethylchlortetracycline.
3 Urea.

Lithium and demethylchlortetracycline interfere with cellular action of ADH and produce nephrogenic DI in man. They interfere with the normal ADH mediated stimulation of adenyl cyclase to increase cyclic AMP production in the renal tubules. In man demethylchlortetracyline (demeclocycline) 600–1200 mg daily is superior to lithium and may obviate the need for water restriction. This drug has also been used for chronic oedema in cirrhotic patients when it induces a substantial increase in Na and water excretion. The increase in Na excretion is independent of ADH and may be due to interference with the action of aldosterone. The patient should be followed up closely. A rise in blood urea and creatinine due to tetracyclines is not uncommon since they have a catabolic action but the possibility of a nephrotoxic effect should always be kept in mind. Oral urea 30–60 g daily can produce a sufficient osmotic diuresis to correct hyponatraemia without major side effects for up to 270 days. 40–80 mg frusemide and 3–6 g NaCl tablets with K supplements or a potassium conserving diuretic may also be needed.

Ethanol acts centrally to suppress ADH release from a neurohypophyseal source of ADH rather than an ectopic source but has obvious limitations for long term treatment.

The ideal drug for the treatment of this condition would be a *competitive inhibitor of the action of ADH*. These agents are becoming available and antagonize the *in vivo* action of vasopressin without changing blood pressure, glomerular filtration rate or solute excretion and have been used successfully in animals. However, these drugs also antagonize the vasoconstrictor effects of vasopressin and where this is raised in conditions of decreased effective circulating blood volume then hypotension would be expected from use of the antagonist. Hyponatraemia can be corrected rapidly but this has been associated with development of central pontine myelinosis. These drugs remain to be fully evaluated.

In summary, water restriction is the treatment of choice for this syndrome but in the acutely symptomatic or the chronic syndrome one of these alternative therapies should be considered.

As this syndrome includes such a variety of aetiological factors in which the increase in ADH secretion is not always inappropriate (it is a normal finding in stress) and as the pathophysiology is not totally clear some authors suggest abandoning this nomenclature and describing the clinical circumstances for example hyponatraemia associated with carcinoma of the bronchus.

DISORDERS OF OSMOREGULATION

These have been referred to as they occurred in this chapter and Chapter 7, but will be considered here for the sake of completeness.

Hypo-osmolality

This is due either to:
1 Water overload, or
2 Solute loss.
Symptoms occur with a plasma osmolality less than 250 mosmol kg^{-1} or plasma Na less than 120 mmol l^{-1} and include headache, anorexia, nausea and vomiting, irritability, restlessness and weakness. At a plasma Na less than 110 mmol l^{-1} confusion and convulsions develop. Death occurs when plasma osmolality is less than 210 mosmol kg^{-1}. Cardiovascular function is preserved but if plasma osmolality is very low pulmonary oedema may occur. This is one circumstance of hyponatraemia where measurement of osmolality can be used to assess whether the low Na is due to a reduction in plasma water content provided permeability of solutes is taken into account. Calculation of effective osmolality assumes that plasma water fraction is not significantly different from normal and there is no gross excess of mannitol. Management of water excess is discussed above and solute loss in Chapter 7.

Hyperosmolality

This is due to either:
1 Increase in solute, or
2 Water loss.

Total ECF solute in a 70 kg man is less than 4000 mosmol. The solute load from hypertonic solutions may be enormous as in the following examples:

Dextrose 50%   2525 mosmol kg^{-1}
Mannitol 20%   1099 mosmol kg^{-1}

Impermeate solutes such as mannitol and sorbitol increase plasma osmolar load but, of greater importance, they induce hypotonic fluid losses by producing an osmotic diuresis. More water is excreted in the urine than sodium thereby inducing progressive hypernatraemia. Hyperosmolality due to *permeate* solutes can exist without hypertonicity and is not uncommon in excess alcohol ingestion. Such a situation can be detected by the difference between calculated osmolality and measured osmolality (p. 10). An osmolar gap of 30–70 mosmol kg^{-1} is likely. This hyperosmolality without hypertonicity does not change ICF volume because osmotic equilibrium can occur by solute movement into cells. Causes of water loss have been discussed above.

Table 6.3 Cause of solute overload.

Solute	Cause
Glucose	Hyperosmolar non-ketotic coma
	Intravenous feeding
Urea	Renal failure
Sodium	Intravenous feeding
	Infant feeding
	Saline enema
Mannitol, glycerol } Sucrose, sorbitol } Fructose }	Overadministration
Alcohol	Rare

The osmolality of enteral feeds for the newborn may be remarkably high and lead to a variety of adverse effects. Even administration of fluids such as lucozade in children with gastroenteritis can induce hypernatraemia because of its high osmolality (710 mosmol kg^{-1}).

Symptoms of hyperosmolality include thirst, tachycardia, hypotension and hyperthermia. Cerebral dehydration causes confusion, hallucinations and coma with convulsions. Symptoms occur at a plasma osmolality greater than 320 mosmol kg^{-1}. Hyperosmolality implies cellular dehydration. In this respect brain cells in the human appear to be unique in that they generate idiogenic osmoles during states of dehydration and hyperosmolality. In experimental hypernatraemia brain volume is substantially reduced after one hour but after 7 days it has returned to its normal size in the presence of continued hypernatraemia. Where therefore does this new intracellular solute to increase brain cell volume come from? Half apparently is ECF Na, K and Cl, which move into the cell. The other half is new solute generated within the cell, at least some of which is amino acid. In experimental hyperglycaemia it takes only 4 hours for the brain volume to return to normal and for idiogenic osmoles to be identified. If dehydration is excessively rapid there is no time for development of these idiogenic osmoles. Equally correction of hyperosmolality overrapidly may result in intracellular oedema since these idiogenic osmoles cannot be removed or inactivated fast enough.

Sudden ECF hyperosmolality increases ECF volume and may produce pulmonary oedema. Usually, hyperosmolality is due to water and some salt loss so that a combination of ECF and some ICF depletion occurs. Pure water loss is relatively rare. If it is profound enough to produce shock it will cause severe hyperna-

Table 6.4 Abnormalities of osmolality.

Diagnosis	Test	Measurement	Reading (mosmol kg^{-1})
Normal		Plasma osmolality	280–295
	Water deprivation (after 8 hours)	Plasma osmolality	300
		Urine osmolality	600
Diabetes insipidus	Water deprivation (after 8 hours)	Plasma osmolality	300
		Urine osmolality	270
		U:P osmolality ratio	0.9–1.0
Inappropriate ADH secretion		Plasma Na concentration	125 mmol l^{-1}
		Plasma osmolality	270

traemia with a Na concentration greater than 170 mmol l^{-1}. If plasma Na is less than this then in the presence of hypotension salt and water has been lost. In other words hypotonic loss has occurred. Infusion of hyperosmolar urea to reduce brain volume for neurosurgery can result in hypotension due to reduction in ionized Ca.

The combination of ECF volume depletion and hyperosmolality both stimulate ADH secretion. Water retention occurs to restore osmolality towards normal. Thirst is also stimulated and increases water intake (without salt) in the presence of the increase in ADH which may rarely actually produce hypo-osmolality.

Treatment

Treatment is to replace any losses with the appropriate fluid over 2–3 days to avoid isotonic water intoxication. These aspects are discussed elsewhere. Hyperglycaemia as a cause of hyper-osmolality is discussed under diabetes in Chapter 14.

Chapter 7
Abnormal Sodium
and Chloride Balance

Acute disturbances of electrolyte balance produce nonspecific symptoms and hence are difficult to detect clinically. Their presence is often suspected when a 'routine urea and electrolyte' request returns an abnormal result. In view of this it is essential to fully evaluate a patient clinically when such an abnormal result is found.

Clinical problems of salt balance are exemplified by changes in extracellular fluid (ECF) volume. Plasma Na concentration only reflects the ratio between total ECF Na and total ECF water and abnormalities of plasma Na can be due to changes in either or both these compartments.

Normally plasma Na is maintained within narrow limits ($\pm 2\%$) by varying water intake and renal Na excretion. Following salt loss and reduced plasma Na, thirst and ADH secretion are inhibited and initially plasma Na is preserved but at the expense of ECF volume. The kidney can reduce Na excretion to as little as 5 mmol l^{-1} in severe Na retention and following a salt load can excrete as much as 250 mmol l^{-1} Na.

HYPONATRAEMIA

An enormous number of pathological states can produce hyponatraemia. The effects of such a state will vary with the speed of onset of the electrolyte abnormality. Severe hyponatraemia of rapid onset may be fatal, producing cerebral oedema within hours, rapidly followed by coma and convulsions. This is a situation with a 50% mortality. Slowly developing hyponatraemia, however, may produce minimal effects, and a plasma Na as low as 89 mmol l^{-1} in a conscious patient has been recorded.

Definition of hyponatraemia

Plasma Na concentration less than 130 mmol l^{-1}.

It is essential to distinguish between true hyponatraemia and pseudohyponatraemia. A spuriously low Na (pseudohyponatraemia) may occur when a large amount of solid matter is present in the plasma, chiefly as lipids or protein. Electrolytes are present only in the aqueous phase of plasma but their concentration is measured and expressed as mmol per total volume of plasma. The lipid volume of plasma is expanded in conditions of diabetic ketosis, nephrotic syndrome and during lipid infusions. If, for example, 20% of the plasma volume is occupied by lipid and the measured concentration of Na is 120 mmol l^{-1} then the true plasma Na is 150 mmol l^{-1}. All obviously lipaemic plasma should be ultracentrifuged before analysis. Measurement of plasma concentrations of lipids is not routine but there are ways of ensuring that pseudohyponatraemia does not exist.

Methods of assessing hyponatraemia

1 Measurement of plasma osmolality. A calculated plasma osmolality is valueless since it depends on the apparent plasma Na.
2 Plasma levels of other electrolytes such as K, Cl, and HCO_3 will also be low in hyperlipidaemia.
3 Urine measurements may be of some value provided there is no renal salt losing syndrome. The major diagnostic value of urinary sodium is in patients with volume depletion, acute oliguria, severe stress response to trauma and hyponatraemia. This is discussed further later in this chapter.
4 The use of an ion selective electrode (ISE) electrolyte analyser which does not dilute plasma prior to measurement, for example the Nova Biomedical 'Nova I'. This machine measures the activity of ions in plasma water and therefore is independent of lipids and proteins which reduce the plasma water space.

Symptoms of hyponatraemia

Symptoms are rare until plasma Na is less than 120 mmol l^{-1}, but include:
   Headache.

Muscle cramps and weakness.
Thirst.
Nausea.
Agitation.
Anorexia.
Disorientation.
Apathy.
Lethargy.

As plasma Na continues to fall to <110 mmol l^{-1} then drowsiness progresses to coma with convulsions. Tendon reflexes may then be diminished and rigidity, extensor plantar responses and pseudo-bulbar palsy may be found.

Hyponatraemia is a laboratory abnormality. The way in which this is related to the clinical state of the patient and his subsequent management depends on the state of the ECF volume. Hyponatraemia can occur with a diminished, expanded or normal ECF volume.

Hyponatraemia with decreased ECF volume

Causes

EXCESSIVE Na LOSS

From the gastrointestinal tract.
   Nausea.
   Vomiting.
   Diarrhoea.
   Fistulae.
   Gastric aspiration.
   Hidden losses in ileus and intestinal obstruction.

From the kidney.
   Salt losing conditions.
   Excessive diuretic therapy.
   Polycystic renal disease.
   Medullary cystic disease.
   Nephrocalcinosis.
   Analgesic nephropathy.

Interstitial nephritis.

Post-obstructive diuresis as acute situation.

The diuretic phase of an acute tubular necrosis.

Administration of Mannitol or Dextran.

Hyperglycaemia (osmotic diuretic effect).

Adrenal insufficiency.

Severe alkalosis. In these circumstances increased urinary loss of bicarbonate necessitates an accompanying cation (Na).

Skin.

Severe sweating.

Burns.

Exfoliative dermatitis.

Exudates.

Peritonitis.

Pancreatitis.

Excessive removal of serous effusions, for example ascites.

Trauma.

Sequestration of ECF occurs in severely traumatized limbs. If these fluid 'losses' are replaced with hypotonic fluid, hyponatraemia will occur.

These salt losses are almost always associated with water loss as well and variable potassium loss. The diagnosis of such salt losses is not especially difficult but if another solute such as glucose is present in excess, this will tend to hold water in ECF and the severity of the situation may be underestimated. When hyperglycaemia is treated, the degree of salt deficiency and ECF depletion may be unmasked. It is therefore very important to measure glucose in initial assessment of hyponatraemia.

DECREASED INTAKE

This is rare and usually iatrogenic such as occurs when acute losses due to sweating or from the gastrointestinal tract are replaced with 5% dextrose only.

When hyponatraemia occurs with a reduced ECF volume the following clinical signs occur:

Postural hypotension.

Low pulse volume.

Reduced central venous pressure.
Loss of skin turgor.
Reduced eyeball tension.
Dry skin and mucous membranes.
Peripheral vasoconstriction.
Oliguria.

If the losses are not replaced the situation progresses to the shock state and metabolic acidosis will accompany these clinical signs.

The salt loss is initially from the plasma volume and ECF compartment, tending to render it hypo-osmolar, so that water passes into cells to maintain osmotic equilibrium. This produces the severe reduction in ECF volume required to produce the clinical signs. In order to maintain the cardiac output the following changes occur:

1   Pulse rate and peripheral resistance rise due to increased sympathetic nervous system activity.

2   The renin angiotensin system is activated by one of the following mechanisms:

(a)   Sympathetic nervous system
(b)   Catecholamines.
(c)   Decreased Na concentration in the macula densa.

This results in increased angiotensin II production with further vasoconstriction, increased aldosterone production and Na retention.

3   Increased ADH secretion.

Laboratory findings

BLOOD

Haemoglobin, haematocrit, plasma proteins and urea are all raised.

URINE

Osmolality, specific gravity, creatinine and urea are all increased. Urine Na will be decreased to less than 10 mmol l^{-1} unless there is an intrinsic salt losing problem when the Na will be greater than 30 mmol l^{-1}. Even in these circumstances, if the disturbance is severe enough to cause a reduction in renal perfusion then renal Na will again be low.

Estimation of sodium deficit

Most of the total body Na is in the ECF, which is roughly 20% of the body weight. An estimate of the Na deficit may be obtained therefore by the following formula:

Na deficit = (normal plasma Na – measured plasma Na) $\times 0.2 \times$ body wt in kg.

A 70 kg man, in whom the normal plasma Na should be 140 mmol l^{-1} and measured Na is in fact 120 mmol l^{-1}, has a Na deficit of

$$(140 - 120) \times 0.2 \times 70$$
$$= 20 \times 40 = 280 \text{ mmol}$$

Such an estimation, however, is not accurate and must serve as a guide only since it takes no account of shifts in water into and out of the ECF by osmosis which occur with changes in ECF Na concentration in order to maintain isotonicity between body fluid compartments. Most patients who are fluid depleted in fact have mixed salt and water losses. The extent of fluid depletion can be correlated with the clinical circumstances.

Table 7.1 Loss of body fluids as percentage of body weight.

	Adults	Children (6 months to 6 years of age)
Mild	4.0	5.0
Moderate	6.0	7.5
Severe	8.0	10.0

Mild salt and water depletion rarely exhibits signs of cardiovascular system dysfunction but these become apparent as depletion becomes moderate and with severe depletion there will be tachycardia and extreme hypotension.

Management of hyponatraemia with ECF depletion

Losses require correction. In acute circumstances of Na loss it is rarely possible to replace adequately by the oral route, therefore intravenous replacement is required. The replacement fluid should correspond as closely as possible to the fluid lost. For example, Hartmann's solution is appropriate for gastrointestinal losses and plasma for burns. Hypertonic saline should be avoided and most commonly 0.9% NaCl is used. Losses should be replaced slowly since equilibrium across the blood brain barrier (BBB) is slow so that rapid replacement may lead to disequilibration and deterioration in central nervous system signs. Oral supplements may be more valuable in the stable situation of renal salt losing syndromes.

Estimated losses are a guide only and the repletion may have to be modified due to continued losses or for example development of endotoxinaemia. Although this book does not deal with paediatric fluid balance, it is pertinent to mention here that oral therapy can be undertaken in dehydrated infants with the standard WHO (World Health Organization)/UNICEF glucose electrolyte solution. Use of this solution is discussed in Chapter 6.

Hyponatraemia with clinically normal extracellular fluid volume

This is usually due to water retention (see Chapter 6) and total body Na is normal. There may in fact be a small increase in ECF volume but insufficient to be detected clinically. Na handling by the kidney is usually normal and urine Na is greater than 20 mmol l^{-1}. In other words the urinary Na reflects the dietary intake. If ECF depletion occurs for any other reason, the kidney can retain Na. Sustained release of ADH occurs despite low plasma Na concentration. This whole subject is discussed more fully in Chapter 6.

Causes of hyponatraemia with clinically normal ECF volume

1 Excessive ADH.
2 Glucocorticoid deficiency. Addison's disease (those cases already diagnosed are often undertreated with fludrocortisone).

3 Severe hypothyroidism.
4 Diuretics (usually a thiazide diuretic used in treatment of hypertension).
5 Water overload with normal renal function.

Excessive ADH production has been discussed elsewhere. In glucocorticoid deficiency one of the main features is inability to excrete a water load. Replacement of the normal physiological concentrations of steroid restores the kidney's ability to excrete dilute urine. The action of steroid in these circumstances may be a direct effect on the osmoreceptors in the hypothalamus or a permissive action on collecting ducts, making them sensitive to ADH and impermeable to water.

Diuretics usually cause isosmotic Na excretion. Hyponatraemia may occur, however, in the following circumstances:

(a)   ECF volume contraction which stimulates ADH secretion.
(b)   Impaired urinary dilution when diuretics block Na reabsorption from cortical diluting segments.
(c)   Potassium deficiency.
(d)   Thirst resulting in increased hypotonic oral fluid intake.
(e)   Overdiuresis, resulting in excess salt and water loss and its replacement by hypotonic fluid.
(f)   Possible increased tubular sensitivity to ADH.

Synergism between diuretics and chlorpropamide or trimethoprim worsens hyponatraemia.

Infusion of Mg in diuretic induced hyponatraemia in severe congestive cardiac failure results in a rise in plasma Na probably due to the effect of Mg on membrane ATPase.

BEER DRINKERS' HYPONATRAEMIA

Consumption of 5 litres or more of beer daily is increasingly being recognized as a cause of hyponatraemia. Although beer is hyperosmolar because of its alcohol content it has a low Na content (2 mmol l^{-1}) and if the diluting ability of the kidney (30 mosmol kg^{-1}) is exceeded, then effectively water retention occurs. Beer incidentally also increases prolactin secretion.

LEGIONNAIRE'S DISEASE

LEGIONNAIRE'S DISEASE

Hyponatraemia in Legionnaire's disease is likely to be multifactorial and an individual approach to cause and treatment is required.

HYPOALBUMINAEMIC HYPONATRAEMIA

Attention has recently been drawn to this syndrome which has some features of inappropriate ADH secretion with low plasma osmolality and inappropriately high urine osmolality but which responds to infusion of albumin by reduction of ADH. Plasma albumin is an important determinant of plasma volume and oncotic pressure and a low circulating volume is an appropriate stimulus for ADH release.

Management

Management is that of the underlying condition (see Chapter 6). Usually fluid restriction corrects hyponatraemia, but the value of this in an asymptomatic patient with minimal expansion of ECF volume is questionable.

Hyponatraemia with expansion of ECF volume

In this situation total ECF Na may be considerably increased but ECF water increases proportionately more resulting in hyponatraemia. Retention of Na is associated with the inability to excrete water normally. Oedema usually accompanies these situations.

Causes

1 Cardiac failure.
2 Renal failure.
(a) Acute renal failure with oliguria.
(b) Chronic renal failure.
3 Nephrotic syndrome.
4 Hepatic insufficiency.
5 Trauma.
Except in renal failure, urine Na is usually less than 10 mmol l^{-1}.

Cardiac failure

As cardiac failure worsens, decompensation occurs with a fall in
cardiac output. This results in a reduced distending pressure within
the systemic arterial vasculature, which is perceived as a reduction
in effective arterial blood volume. Such a patient behaves as
though he were volume depleted and retains salt and water avidly,
due to the following mechanisms:

(a)   Reduction in renal blood flow and GFR.
(b)   Increased tubular reabsorption of Na and water.
(c)   Redistribution of renal blood flow.

Recently it has been shown that patients with severe congestive
cardiac failure have increased atrial natriuretic factor (ANF),
which is a peptide. This correlates positively with wedge pressure.
These atrial peptides inhibit endogenous vasoconstrictors and
reduce aldosterone synthesis. The increased ANF is secondary to
raised atrial pressure and plasma ANF could be used as an index of
severity of congestive cardiac failure. High ANF is also found in
patients with supraventricular tachycardia (presumably due to
atrial distention) when it may account for the polyuria. Volume
overloaded children in chronic renal failure and adults with essential hypertension also have raised ANF. In patients with treated
heart failure hyponatraemia correlates inversely with plasma renin
activity and this may be used as a marker to identify those who
may respond to converting enzyme inhibitor.

REDUCTION IN RENAL BLOOD FLOW AND GFR

Reduction in arterial blood pressure is sensed by baroreceptors in
the carotid sinus and elsewhere. This results in increased sympathetic nervous system activity which leads to peripheral and renal
vasoconstriction, resulting in decreased renal blood flow. GFR is
reduced proportionately rather less because constriction affects
mainly efferent arterioles. The renin angiotensin system is stimulated due to reduced effective blood volume and increased sympathetic activity. This results in increased circulating and intrarenal angiotensin II, further renal vasoconstriction and reduction
in renal blood flow and GFR.

INCREASED TUBULAR REABSORPTION OF SALT AND WATER

This is the major factor in increasing Na and water retention in congestive cardiac failure. Precise mechanisms remain to be elucidated. Efferent arteriolar constriction reduces peritubular capillary hydrostatic pressure. This, in conjunction with an increase in peritubular capillary osmotic pressure, which occurs commonly in congestive cardiac failure, would enhance proximal tubular reabsorption. Micropuncture studies have demonstrated increased Na reabsorption in the loop of Henle. In addition, aldosterone secretion is increased, presumably due to stimulation by elevated levels of angiontensin II. This will enhance distal tubular Na reabsorption. Other hormones which may play a part in salt and water retention are:

Natriuretic hormone.
Catecholamines.
Prostaglandins.
Kinins.
ADH.

This aspect has been discussed more fully in Chapter 2.

REDISTRIBUTION OF INTRARENAL BLOOD FLOW

In congestive cardiac failure xenon washout studies and autoradiographic techniques have shown a redistribution of blood flow from the superficial to the deep cortex within the kidney. Deeper cortical nephrons are said to have long loops of Henle and a greater capacity to reabsorb Na.

MANAGEMENT

1 If cardiac output (CO) is reduced, measures should be taken to improve it; these will include:
(a) Appropriate treatment of cardiac arrythmias.
(b) Inotropes, such as digitalis, dopamine.
2 If cardiac output is normal then treatment should be directed to the underlying cause, for example:
(a) Correction of anaemia.
(b) Repair of arteriovenous fistulae.

(c)   Treatment of thyrotoxicosis.

(d)   Treatment of thiamine deficiency.

3 If these measures fail, salt and water restriction are appropriate.

4 Diuretics. If the cardiac failure is mild, thiazides diuretics with potassium supplements may be sufficient. In more severe degrees of congestive cardiac failure, more powerful diuretics such as frusemide (a loop diuretic) perhaps with a distally acting potassium conserving diuretic such as amiloride or spironolactone are more suitable.

Renal failure

If a patient in this group has a urine Na greater than 20 mmol l^{-1} then acute or chronic renal failure causing renal salt loss is usually the problem. Hyponatraemia may then be compounded by iatrogenic administration of intravenous dextrose. Renal disease is discussed more fully in Chapter 13.

Nephrotic syndrome

Expansion of ECF volume may be particularly marked in the nephrotic syndrome, which is characterized by severe oedema, with heavy loss of protein in the urine. This protein loss consists mainly of albumin and the resultant hypoalbuminaemia and reduction in circulating plasma volume or colloid osmotic pressure are potent stimuli for ADH production. Low plasma colloid osmotic pressure favours loss of water and salt from intravascular to interstitial compartments and hence oedema occurs. Renal retention of salt and water occurs but this is soon lost into interstitial fluid because of the persistently low plasma colloid osmotic pressure.

MANAGEMENT

Management of nephrotic syndrome consists of treating any glomerular disease, for example, using steroids in minimal lesion glomerular disease, administering a high protein diet and restricting salt and water. When diuretic therapy is started, a brisk diuresis may occur, producing a further reduction in circulating blood volume and cardiovascular collapse. Concurrent administration of albumin often enhances the diuresis. In the long standing cases large doses of a potent loop diuretic may be required.

Hepatic cirrhosis

Ascites is initiated by transudation of fluid into the peritoneal cavity from congested lymphatics on the surface of the liver. If this loss of fluid is marked, it leads to intravascular volume depletion and stimulation of the mechanisms for salt and water retention. Aldosterone secretion is raised and in severe liver impairment, its metabolism is reduced. Hypoalbuminaemia is usual in severe cirrhosis. Continuing salt and water retention is potentiated by:
1 Hepatic venous obstruction.
2 Inferior vena caval obstruction, producing a fall in venous return and cardiac output.
3 Myocardial disease due to alcohol.
4 Poor nutritional state.
5 High output cardiac failure due to abnormal vascular anastomoses within liver and lungs.
Treatment with propranolol in patients with chronic liver disease does seem to reduce the increase in total body water and sodium which occurs in the untreated patients.

MANAGEMENT

Management is largely that of the primary liver disease and includes improvement in nutrition, with salt and water restriction.
Diuretics. Hypokalaemia is common in hepatic cirrhosis so it is important to use a potassium conserving diuretic. If potent loop diuretics are required, potassium supplements will also be necessary. Often these patients are resistent to diuretics and although water restriction may help, the use of urea to increase the diuresis has also been successful, presumably by inducing an osmotic diuresis. This is contraindicated, however, if renal function is impaired. If these measures fail to control ascites and paracentesis is required for the patient's comfort or to facilitate respiratory exchange, then retransfusion of the high protein fluid is a useful manoeuvre. Co-existent cardiac failure should be treated along the usual lines.

Trauma

Following trauma and surgery and in hypercatabolism, tissue breakdown occurs releasing water which will expand total body water. As this water is freely diffusible between all compartments, it will produce hyponatraemia.

Oedema

Sodium overload is associated with expansion of the intravascular and interstitial fluid compartments with raised central venous pressure and oedema. Causes for oedema include acute glomerular nephritis, nephrotic syndrome, renal failure, congestive cardiac failure, cirrhosis, pregnancy and pre-eclamptic toxaemia. Plasma sodium may be high, normal or low. Treatment is that of the underlying cause but bed rest, dietary restriction, diuretics and ultrafiltration techniques (see Chapter 13) may be required. Local sodium retention and oedema can be caused by venous or lymphatic obstruction.

The oedema and hypertension associated with pre-eclamptic toxaemia of pregnancy is associated with a reduction in plasma volume. The osmolar properties of ISF albumin play a key role in development of physiological and pathological oedema of pregnancy.

Idiopathic oedema

In this situation, both salt and water retention may occur. Idiopathic oedema is a condition that occurs principally in females. Fluid retention occurs predominantly during the day while the patient is ambulent, with diuresis at night. The aetiology is uncertain but may be due to increased permeability of the capillary wall, which increases transudation of salt and water into the interstitium, producing intravascular depletion. Management includes reassurance, salt and water restriction, rest during the day and diuretics which if overused will reduce intravascular volume and perpetuate the oedema producing mechanisms.

The syndrome of cyclical oedema may be due to hypothalamic disorder.

Hyponatraemia and sick cells

In many very ill patients plasma Na is low due to widespread increase in cell membrane permeability. This results in flow of Na ions along their concentration gradient from ECF into cells and K flow in the reverse direction to increase plasma K. This hyponatraemia worsens as the patient's condition deteriorates and improves with clinical improvement. Cell membrane permeability is known to be increased by hypoxia, substrate depletion (decreased ATP), metabolic inhibitors and endotoxin.

Another possible mechanism for this hyponatraemia is reduction in ICF osmolality which occurs in the very ill patient and results in water leaving the cell to render ECF hypo-osmolar. ADH secretion results in water retention and further ECF dilution. The increased membrane permeability may allow Ca ions into hypothalamic osmoreceptors with nonspecific release of ADH. Administration of Na in the sick cell syndrome is dangerous and will further increase K loss from the cell.

A further alternative is that impaired excretion of water occurs as, for example, in liver failure and that this is associated with increased ADH levels. A similar association occurs with congestive cardiac failure and other situations of reduced effective plasma volume which is sensed by baroreceptors and non-osmotic stimulation of ADH occurs. In support of this is the absent osmolar gap in sick cell syndrome.

The sick cell concept requires that a sufficient percentage of cells fail metabolically to sustain their normal content of *non*-diffusible solutes so that their osmolality falls. Water will pass out of the cell producing reduction in ECF osmolality and a fall in plasma sodium. The cells must be so abnormally permeable that they leak normally non-diffusible solute for this concept to hold water.

In these circumstances measures which increase the activity of the Na pump may be valuable. These measures include:

1   Administration of glucose and insulin with added K depending on the plasma K concentration.
2   Steroid administration.

In these very sick patients, the circulation may be unstable and a plasma expander may be necessary. Human plasma protein fraction (HPPF) may be valuable, but unfortunately has a high Na content and salt-free albumin in these circumstances may be preferred. This subject is discussed more fully in Chapter 11.

Aspects of management of hyponatraemia have been discussed under individual categories. Investigations include plasma urea and electrolytes, creatinine, glucose, cortisol, proteins and osmolality and urine osmolality and electrolyte content. The patient should be weighed daily. The presence of a low plasma Na requires more active treatment if the fall in Na is acute. Giving Na for hyponatraemia diagnosed from a pathology report is illogical and potentially dangerous. A laboratory result should NEVER be treated without a full history and examination of the patient. Formulae which calculate the Na deficit should be used as guidelines only and accurate, repeated measurement of plasma and urinary electrolytes are mandatory with regular review of the clinical state of the patient. Although salt replacement undertaken with care leads to relatively few problems, acute psychosis and hypertensive crises are documented. Disturbances which develop over a prolonged period seldom require acute treatment, only that directed at the underlying disease. Patients in renal failure may require dialysis to remove excess salt and water. This is discussed more fully in Chapter 13.

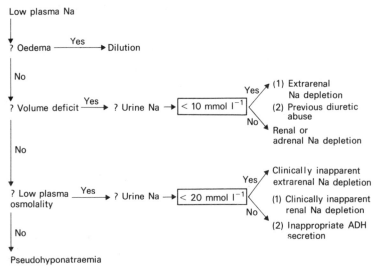

Fig. 7.1 Summary of causes of hyponatraemia.

Chronic sodium depletion occurs when cardiac output is reduced with the renin-angiotensin system actively maintaining mean arterial pressure. Secondary hyperaldosteronism occurs. Hepatic blood flow falls due to increased angiotensin II which increases splanchnic vascular resistance. This is reversed with saralasin.

HYPERNATRAEMIA

Definition

Plasma Na concentration greater than 140 mmol l^{-1}. Such a situation implies a deficiency of water for the total body solute. These problems have already been discussed in Chapter 6. A high oral Na intake is very unlikely to result in hypernatraemia because of intense stimulation of thirst. Patients with Cushing's syndrome and hyperaldosteronism, despite Na retention, rarely present with hypernatraemia and the accompanying hypokalaemic alkalosis suggest the diagnosis. Hypernatraemia has resulted from absorption of 30% saline used to irrigate after removal of an hydatid cyst. Hypernatraemia with acidosis occurs in sodium valproate poisoning.

Excessive protein intake in the course of nasogastric feeding may increase urea production and result in osmotic diuresis. Such an osmotic diuresis induced therapeutically with mannitol or accidentally by hyperglycaemia is an important cause of hypernatraemia. Prolonged iatrogenic parenteral infusion of saline especially hypertonic $NaHCO_3$ may cause hypernatraemia.

In a recent study of severe hypernatraemia in adults with plasma Na greater than 154 mmol l^{-1} the commonest causes were diabetes mellitus and intracranial disorders. Diabetics may develop a hyperosmolar, nonketotic state with hypernatraemia despite treatment with hypotonic 0.45% saline. Intracranial disorders may be accompanied by abnormal production of ADH and frank DI is not uncommon. When urine Na excretion is low, infusion of 0.9% saline may make the hypernatraemia worse.

Management

This consists of encouraging oral water intake or administering 5% dextrose parenterally slowly.

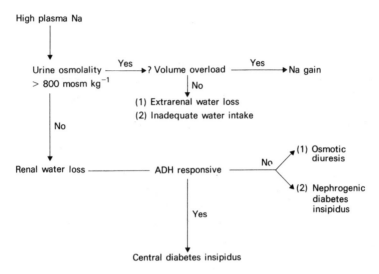

High plasma Na

Urine osmolality —Yes→ ? Volume overload —Yes→ Na gain
> 800 mosm kg^{-1} │ No

(1) Extrarenal water loss
(2) Inadequate water intake

No

Renal water loss ——————— ADH responsive ———— No → (1) Osmotic diuresis

▶ (2) Nephrogenic diabetes insipidus

Yes

Central diabetes insipidus

Fig. 7.2 Summary of causes of hypernatraemia

Sodium retention

Many drugs cause sodium and water retention:
Corticosteroids, mineralocorticoids, ACTH.
Oestrogens.
Anti-inflammatory drugs.
Carbenoxolone, liquorice.
Arteriolar vasodilators.
Hypotensive agents.
Psychotropic drugs.
Drugs with a high sodium content.
beta lactam antibiotics
antacids
i.v. fluids
X-ray contrast media
Management includes withdrawal of the drug, sodium restriction and perhaps diuretics.

ABNORMALITIES OF CHLORIDE BALANCE

These closely mirror disturbances of Na and occur in the same circumstances.

Hypochloraemia

Causes

Increased secretion or loss of gastric juice, for example in pyloric stenosis and prolonged nasogastric suction.

Increased renal excretion, diuretics.

Disorders of Na regulation — aldosteronism.

Dilution.

Actual hyponatraemia.

Increased bicarbonate retention.

Treatment is that of the underlying cause. If the plasma chloride is greater than 90 mmol l^{-1} then the loss does not need replacing. An estimate of the chloride deficit may be calculated as follows:

Chloride deficit = [Normal plasma chloride (100 mmol l^{-1}) − measured plasma chloride] × 0.2 × body weight in kg

This is along the lines of estimating Na deficit (see above). If a true chloride deficit exists then it should be replaced with 0.9% NaCl.

Hyperchloraemia

Causes

Increased intake or administration.

Decreased production of bicarbonate.

Respiratory alkalosis.

Decreased excretion by the kidney unable to produce bicarbonate for example, in renal tubular acidosis.

Dehydration.

Excessive absorption from the gastrointestinal tract following ureteroileostomies.

Treatment is again that of the underlying cause and almost always correction of Na abnormality is of primary importance and may in itself correct the abnormality of chloride.

VALUE OF URINE SODIUM AND CHLORIDE MEASUREMENTS

A random urine urea and electrolyte estimation is valueless.

Urine sodium

Measurement of urine Na may be of value when the diagnosis is otherwise not apparent in states of:
1 Volume depletion.
2 Acute oliguria.
3 Hyponatraemia.

Volume depletion

The body is very efficient in conserving Na when intake is reduced. This is a mechanism which helps to preserve ECF volume. It is mediated by some reduction in GFR and an increase in tubular Na reabsorption. If a patient is volume depleted therefore, the virtual absence of Na in the urine indicates an extrarenal Na loss. If, however, the urine does contain appreciable Na, which is a much rarer occurrence, then the diagnosis is one of renal salt wasting. It is very important in this context to ensure that the patient is not receiving diuretics or mannitol (or any other osmotic diuretic such as glucose). In addition to renal salt wasting a high Na loss will also occur in adrenal insufficiency when the tubular ability to reabsorb Na is impaired. Medullary cystic disease results in a very severe impairment of renal salt conservation, requiring up to 15 g of NaCl daily to maintain Na balance. During the diuretic phase of acute tubular necrosis similar large losses of salt may occur.

Urine Na less than 10 mmol l^{-1} occurs in extrarenal salt loss.

Urine Na greater than 10 mmol l^{-1} occurs in renal salt wasting or adrenal insufficiency.

Acute oliguria

In an oliguric patient it is very important to distinguish between prerenal causes of oliguria and established acute tubular necrosis. In this context clinical evaluation of the patient and assessment of the state of the circulation using central venous pressure measurement are of paramount importance. In volume depletion alone urine Na should be low, but in acute tubular necrosis and some other renal disorders such as obstructive uropathy urine Na is usually greater than 35 mmol l^{-1}.

Urine Na less than 10 mmol l^{-1} implies a prerenal cause.

Urine Na greater than 35 mmol l^{-1} implies acute tubular necrosis.

Hyponatraemia

In hyponatraemia urine Na may be of value to relate dietary intake to renal loss when the patient is hypo-osmolar. If urine Na is low in the presence of an unrestricted Na intake then the Na is being retained either to expand a depleted ECF or due to Na retention in oedematous states. Hyponatraemia is due to inadequate excretion of free water. If, however, urine Na is equal to or in excess of dietary intake then inappropriate ADH secretion or adrenal insufficiency are the likely diagnoses.

Urine Na less than 10 mmol l^{-1} occurs in volume depletion or oedematous states.

Urine Na equal to or greater than dietary intake occurs in inappropriate ADH secretion or adrenal insufficiency.

Urine chloride

This is of limited value in patients with metabolic alkalosis which may be due to chloride depletion. A urine chloride of less than 10 mmol l^{-1} implies that the patient should respond to administered chloride. In hyperadrenocorticism, on the other hand, renal bicarbonate reabsorption is stimulated and these patients remain alkalotic despite chloride administration because this ion is lost in the urine.

SALT AND HYPERTENSION

There is at present considerable controversy over the relation of Na intake to blood pressure and the value of Na restriction in the treatment of hypertension.

Very low Na intake occurs on the Papuan highlanders of New Guinea and other hunter gatherers (less than 30 mmol Na daily). In these groups blood pressure does not rise with age and essential hypertension is unknown. However, in most populations blood pressure does rise with age. Exchangeable sodium is positively correlated with blood pressure (BP), most closely in the elderly. Exchangeable potassium is inversely related to BP especially in younger patients. This leads to suggestions that development of hypertension in its early stage was due to a K related process. Then it was postulated that a renal lesion developed perhaps as a result of the hypertension and this was characterized by a resetting of the BP. Natriuresis was then susceptible to increased dietary Na.

Very high Na intakes have been given experimentally (600–800 mmol daily). These high levels can cause blood pressure to rise. The hypertensive response elicited by acute hypertonic saline is due to vasoconstriction mediated in part by vasopressin and in part by the sympathetic system. Experimentally a more modest increase in Na intake by 100–200 mmol daily causes blood pressure to rise in normal volunteers and in patients with mild essential hypertension. However, it must be said that it is difficult to accurately evaluate salt intake and output. Most populations have a daily average Na intake of 90–120 mmol and the blood pressure rises with age. This, however, is not pathological.

It is not possible to correlate an individual's blood pressure with either his Na intake or his urinary Na. In the past very low Na diets (10 mmol daily) such as the Kempner rice–fruit diet were shown to reduce blood pressure in a hypertensive patient. Recent studies of moderate salt restriction by not adding salt to food after cooking have shown a reduction in blood pressure with 100 mmol or less Na per day especially if this is accompanied by weight loss or beta blocking therapy. Similarly a high K intake may be associated with a fall in BP but only if Na restriction cannot be achieved. High K intake improves compliance on a low salt diet, promotes Na excretion, prevents the increased catecholamine release induced by a low salt diet and increases the sensitivity of the baroreceptor reflex. There is, however, a wide variation in response.

It may be therefore that dietary exposure to high Na intake is an important factor in the development of essential hypertension in susceptible individuals, perhaps those with reduced responsiveness of the renin-angiotensin system. In this context it would be of value if the Na content of foodstuffs was printed on the wrapper. Controlled studies of salt restriction versus placebo therapy in mild hypertension show equal reduction of blood pressure in both groups after one year which probably reflects the increases monitoring and consultation in the placebo group.

Some recent evidence suggests that the rise in blood pressure in essential hypertension is due to an increase in circulating sodium transport inhibitor. In patients with essential hypertension and their parents salt loading and potassium depletion results in low Na/K net flux in erythrocytes due to a defect in Na/K transport. Plasma from hypertensive patients suppresses the sodium pump in cells of normal people as it contains a sodium transport inhibitor with natriuretic properties. It is postulated that this is continuously correcting the underlying tendency of the kidney to retain salt as it is likely in man that the primary abnormality in essential hypertension is in the kidney. This mechanism should keep ECF volume within normal limits.

However, the increased concentration of circulating Na transport inhibitor might increase the tone in arterioles and thereby increase blood pressure. These abnormalities are greater in patients with low renin hypertension where plasma renin does not increase in response to salt restriction. Catecholamines may contribute to the disturbance of membrane cation handling.

Atrial natriuretic peptides are increased in essential hypertension probably as a secondary phenomenon to avoid sodium overload. The increased intracellular Na levels which result from

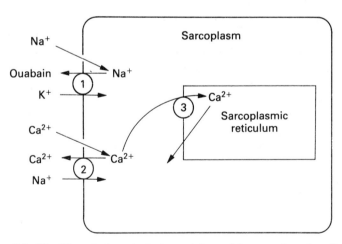

Fig. 7.3 The Blaustein hypothesis for aetiology of hypertension. A sodium–calcium countertransport mechanism (2) is inhibited by increased intracellular sodium and so intracellular calcium accumulates and causes increased reactivity.

the inhibition of sodium pump activity inhibits Ca efflux with an increase in intracellular Ca which increases vascular smooth muscle tone. Verapamil reverses the increased intracellular Ca and corrects hypertension. The Blaustein hypothesis which explains these events is illustrated in Fig. 7.3.

Circulating levels of renal Na/K dependent ATPase are 25 times higher in healthy subjects on a high salt diet than a low salt diet suggesting that Na excretion is controlled by a natriuretic substance.

In hypertension the kidney fails to generate dopamine in response to a salt load. Dopamine behaves as an intrarenal hormone. A low salt diet (20 mmol) is associated with a low level of dopamine in the urine and with salt loading urine dopamine rises. In hypertension this rise in urine dopamine (and therefore increased natriuresis) does not occur. This may be the cause or a consequence of hypertension but it may lead to either inappropriate vasoconstriction or inefficient Na excretion, both of which may increase arterial blood pressure. There may be specific ANF binding sites in the kidney.

Essential hypertension is probably due therefore to a combination of genetic and environmental factors. Hypertensive patients do have an accelerated natriuresis when given a sodium load. There is a large overlap between normal and abnormal patients in Na/K transport activity and so this is not a useful genetic marker. Young adults with a familial predisposition to hypertension behave similarly to those without such predisposition in having a pressor response to a high Na intake, but peculiarly in showing a depressor response to a high K intake. Although salt restriction is probably of no value in severe hypertension, epidemiological evidence suggests that it will produce a reduction in BP in younger patients with milder hypertension which should reduce the incidence of coronary heart disease and stroke. There are, however, many other dietary factors which influence BP such as intake of carbohydrate, fat, caffeine, alcohol and intake and excretion of Ca. It is for this reason that some authorities believe that a campaign of dietary salt restriction is unnecessary, expensive and may be dangerous.

Many of the drugs used to treat hypertension have an effect on Na balance. Thiazide diuretics among other actions increase urinary Na excretion. With the notable exception of beta-adrenergic blocking drugs many antihypertensive agents increase plasma renin. The thiazide derivative, diazoxide, in addition to producing hyperglycaemia has been shown to produce Na retention by this effect on plasma renin. Minoxidil, which is a powerful peripheral vasodilator causes sodium retention which may precipitate congestive cardiac failure.

Angiotensin II is the most potent vasoconstricting agent known to occur naturally in man. Angiotensin II is an octapeptide formed by the action of converting enzyme on angiotensin I, and this in turn is formed from its precursor by the action of renin on angiotensinogen. In addition to the vasoconstrictor effects of angiotensin II it also releases aldosterone with characteristic effects on Na retention. Angiotensin production may be interrupted at various stages in its pathway.

1 Saralasin is a competitive antagonist of angiotensin II which can be administered intravenously. In a hypertensive patient who normally has a high level of circulating renin, saralasin infusion will produce a reduction in blood pressure. Experimentally in normal subjects infusion of saralsin produces no changes in aldosterone, angiotensin II or blood pressure but after Na depletion saralasin infusion causes a fall in aldosterone. This confirms the role of the renin angiotensin system in increasing aldosterone secretion during Na depletion.

2 Converting enzyme inhibitors.

Captopril inhibits angiotensin I converting enzyme. The effect of this is a reduction in angiotensin II and aldosterone but an increase in renin and angiotensin I. Administration of captopril results in a fall in blood pressure proportional to the pretreatment renin concentration. Other factors such as increased survival of bradykinin or increased production of prostaglandins may also contribute to the hypotensive action. The fall in aldosterone results in excretion of Na but this effect is offset by these effects:

(a) Reduction in the direct natriuretic action of angiotensin II.

(b) Reduced natriuresis due to blood pressure reduction.

Plasma K rises which may be due to the reduction in aldosterone. This makes captopril of theoretical value in treatment of Bartter's syndrome. At present captopril is used for the treatment of otherwise refractory hypertension especially in patients with renovascular hypertension. The drug has exhibited some side effects which may be serious such as immunosuppression, agranulocytosis and neuropathy. Renal failure may be precipitated in patients who have renal artery stenosis as may renal artery occlusion. Cerebral blood flow falls in the elderly. Whenever therapy with captopril is undertaken plasma Na and K must be monitored regularly. Hyponatraemia had been reported during treatment of congestive cardiac failure. One possible mechanism for this is the reduction of aldosterone. The originally recommended doses for captopril were almost certainly too high. Where captopril alone fails to control BP adding diuretics is likely to be successful although a beta-blocker less so. The latter drugs reduce plasma renin activity.

Enalapril is a similar drug introduced more recently. Both it and captopril increase urate excretion by a mechanism that may be coupled to sodium in the renal tubule. These drugs may become the vasodilators of choice in treatment of heart failure.

SODIUM AND THE PREMENSTRUAL SYNDROME

Symptoms are probably due to Na and water retention due to aldosterone in the premenstrual phase. In the postovulatory phase progesterone exerts a natriuretic effect. Treatment should be directed at correcting these effects.

Chapter 8
Abnormal Potassium Balance

It has already been stated that total body K can be considerably depleted before a reduction in plasma K occurs (see Chapter 3). Plasma K is affected by pH and its correction. A fall in pH of 0.1 unit may be expected to increase plasma K by 0.4–1.5 mmol l^{-1}. Plasma K is therefore a poor guide to total body K or intracellular K. However, plasma K is regulated by exchange between cells and ECF as well as renal excretion, so that changes in the concentration of ECF K can to some extent be buffered by uptake or release of K from cells.

HYPOKALAEMIA

Definition

A reduction in plasma K to less than 3.5 mmol l^{-1}.

This level varies a little between laboratories. People on a normal diet take in approximately 80 mmol K daily so that deficiency due to reduced intake is rare and most cases are due to increased losses although tube feeding or parenteral nutrition with inadequate K supplementation may result in hypokalaemia. Occasionally in the elderly with a very poor diet K supplements may be indicated. The K concentrations in gastrointestinal secretion are given in Table A4 (p. 109).

Causes of hypokalaemia

Gastrointestinal losses

Vomiting or nasogastric suction.
Malabsorption.

Diarrhoea, often associated with antibiotic use which disturbs bowel flora.

Laxative abuse.

Ureterosigmoidostomy.

Fistulae.

Villous papilloma of the rectum.

Ion exchange resins.

Renal losses

Hyperaldosteronism.

Diuretic therapy, particularly xipamide

Steroid therapy, Cushing's disease.

Ectopic ACTH syndrome.

Liquorice, carbenoxolone.

Excessive renin secretion.

Congestive cardiac failure (CCF).

Nephrotic syndrome.

Cirrhosis.

Liver failure.

Bartter's syndrome.

Renal tubular acidosis.

Diuretic phase of acute renal failure.

Post obstructive diuresis.

Uncontrolled diabetes mellitus.

In primary hyperaldosteronism the cardinal features are low plasma K and raised plasma and urine concentration of aldosterone with suppressed plasma renin activity and risks of associated hypertension and hypokalaemia. Those patients without a tumour have features which overlap with the normal or hypertensive population. A glucocorticoid suppressible hyperaldosteronism suggests a role for pituitary factors in this abnormality.

Diuretics increase K excretion but this is minimized by a liberal Na intake. Increased ADH or aldosterone augment K losses. After a single dose of frusemide there is precise renal compensation to restore K balance within 15 hours unless other K losing factors operate.

Other

Beta receptor agonists.
Alkalosis.
Insulin induced hypoglycaemia.
Periodic paralysis — familial, thyrotoxic.
Overhydration.
Amphotericin, Carbenicillin, Nifedipine.
Severe megaloblastic anaemia.
Primary malignant disease, especially myeloid leukaemia when the kidney seems unable to conserve K.
Cardiopulmonary bypass. Hypokalaemia is partly due to preoperative treatment with diuretics, and partly to hypothermia, uptake of K into cells with glucose, internal redistribution and urine loss.
Inappropriate ADH. In a recent series of the syndrome of inappropriate ADH secretion due to bronchogenic carcinoma there was little change in plasma K unless additional factors were operative.

Increased circulating adrenaline is associated with hypokalaemia (p. 56) and this mechanism is probably responsible, at least in part, for the hypokalaemia observed during abstinence from alcohol in patients with delerium tremens. Beta blockers, especially non-selective ones reduce this hypokalaemia. Beta 2 agonists are widely used for treatment of premature labour and asthma. Inhaled bronchodilator therapy with beta 2 agonists such as terbutaline may produce profound hypokalaemia which could precipitate ventricular dysrhythmias. This is especially dangerous in myocardial infarction and asthma. Selective beta 2 blockade can abolish the hypokalaemia. These agents are more dangerous if used in combination with a thiazide diuretic. The changes are mediated via beta 2 receptors linked to Na/K ATPase causing K influx. Slow infusion of beta agonists also cause a significant fall in Ca and Mg levels and dose related decreases in plasma PO_4. Although the hypokalaemia responds to therapy with beta blockers this may be contraindicated in some disease states. Lithium induces intracellular hypokalaemia although plasma K is often normal. This effect is exaggerated by diuretics.

Despite the fact that diuretics are often given with K supplements these are often inadequate to prevent hypokalaemia. This is especially likely with the powerful loop diuretics such as frusemide and ethacrynic acid. Nevertheless diuretics are an *infrequent* cause of severe hypokalaemia. Diuretic treatment of uncomplicated hypertension is said not to be associated with significant hypokalaemia, but only a small reduction of 5–10% (or about 200 mmol) in total body K. Hypokalaemia and K depletion is commoner in elderly females and may be partly due to secretive laxative abuse. However, it is wise to measure plasma K at the onset of treatment and occasionally thereafter, especially in patients on digoxin where hypokalaemia or hypercalcaemia enhance digoxin toxicity. Long term diuretics may produce a modest fall in plasma K but this is not usually associated with a fall in total body K content, red blood cell K or exchangeable K. In some situations of severe underlying disease, such as CCF, when secondary hyperaldosteronism may occur, K depletion may be present. In general the treatment of hypertension with thiazide diuretics does not require K supplementation.

Hypokalaemia has also been reported in respiratory insufficiency with a fall in exchangeable K of 16–36%. The discharge from villous papilloma of the rectum may have a K concentration as high as 50 mmol l^{-1}.

Signs and symptoms of K depletion

Muscle weakness — this affects both smooth and skeletal muscle
    — leading to cramps and paralytic ileus.
Loss of tendon reflexes.
Fatigue.
Apathy, sleepiness.
Tachycardia.
Hypotension, postural hypotension.
Mental abnormality, confusional state.
Cardiac arrythmia, cardiac arrest.
Increased sensitivity to digitalis.

K and digitalis compete for myocardial binding sites so that hypokalaemia increases digitalis binding in the heart with symptoms of toxicity. Myopathy can be caused by severe hypokalaemia especially with prolonged use of diuretics, purgatives and amphotericin B. K is usually between 1–2 mmol l^{-1} and accompanied by hypochloraemic alkalosis. It may be associated with euphoria and hallucinations.

Hypokalaemic nephropathy occurs and the kidney is unable to excrete a concentrated urine. This leads to polyuria and nocturia.

In a complex clinical situation K losses should always be measured. 24 hour collections of urine, diarrhoea and other gastrointestinal losses from fistulae, nasogastric suction and other drainage should be made especially if these losses exceed 100 ml daily. The ECG may be some guide to the severity of K losses and must certainly be monitored during acute intravenous K replacement. ECG signs of hypokalaemia include a reduction in the height of the T wave, depression of the ST segment and occasional inversion of T waves. In severe cases a U wave appears and the QT

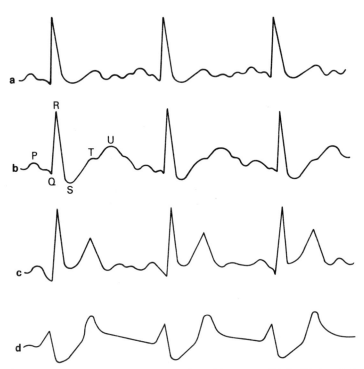

8.1 The ECG in abnormalities of potassium. (a), Normal ECG; (b), hypokalaemia; (c), hyperkalaemia; (d), severe hyperkalaemia.

interval and QRS complex are widened. In chronic K depletion the cardiac muscle is protected against K loss by an adaptive increase in sodium pump density so that cardiac muscle does not show the same reduction in intracellular K concentration as skeletal muscle.

In almost all K losing states coexistent metabolic alkalosis occurs. Some quantitative measure of hypokalaemia is gained from plasma HCO_3 which is increased due to accumulation of H ions inside the K depleted cells. Logically therefore K should be replaced as the chloride salt. In an acute intensive care situation a variable blood sugar and unknown renal function make manipulation of plasma K unpredictable and potentially hazardous.

Estimation of potassium deficit

Since most of the K lost from the body is from ICF the loss occurs from a volume about twice ECF volume or about 40% body weight.

K deficit = (normal K − measured K) × 0.4 body weight (kg)
For example, in a 60 kg female with plasma K 2 mmol l^{-1}:

$$K \text{ deficit} = (4.5\text{–}2.0) \times 0.4 \times 60$$
$$= 2.5 \times 24$$
$$= 60 \text{ mmol}$$

Treatment of hypokalaemia

A decision must first be made as to when replacement becomes necessary. When plasma K is less than 3.0 mmol l^{-1} symptoms such as cardiac arrythmia, digitalis toxicity, myopathy and nephropathy are likely and treatment is required. There is little evidence that the usual degree of hypokalaemia produced by diuretics (3.0–3.5 mmol l^{-1}) is harmful. Patients in whom acute hypokalaemia occurs on admission to hospital with an acute medical condition probably have a shift of K into cells due to adrenaline. This usually reverts to normal after 4 days whether or not oral K supplements are given.

Therefore K should be given to increase plasma K to greater

than 3.0 mmol l^{-1} and to protect patients at special risk, such as the following:

1 Those with severe heart disease on large doses of diuretics.
2 Those on digoxin.
3 Those with severe liver disease since hypokalaemia can precipitate encephalopathy.
4 Those receiving other drugs causing K loss such as corticosteroids or carbenoxolone.

Oral potassium replacement

The following preparations are available:
1 Potassium chloride (KCl). The original formulation led to gastrointestinal ulceration.
2 Enteric coated tablets were subsequently introduced but these are still occasionally associated with small bowel haemorrhage and ulceration.
3 Slow release K (Slow K) tablets.
4 Effervescent K.
5 KCl liquid (Kay Cee L). This is unpalatable unless given through a nasogastric tube, though very useful for rapid correction of hypokalaemia. It is rapidly absorbed from the stomach.
The usual daily dose of K is 24–63 mmol, but up to 150 mmol may be required daily in severe cases.

COMBINED TABLETS

Some manufacturers now produce a combined diuretic with K supplement in a single tablet. Usually the K content is relatively low (8 mmol) and may be quite insufficient for replacement purposes. Patients prefer such combined tablets for convenience and they do ensure that at least some K is taken with the diuretic.

POTASSIUM SPARING DIURETICS

These agents which include amiloride and spironolactone are of value in preventing hypokalaemia in patients at special risk such as those on digoxin but can cause hyperkalaemia in the presence of renal impairment.

HAZARDS OF ORAL K REPLACEMENT

1 Hyperkalaemia. K supplements in renal failure can produce dangerous hyperkalaemia. This can sometimes also be the case with K conserving diuretics such as amiloride used in the presence of impaired renal function. In the elderly subcutaneous infusion has been used to reduce the complication of acute hyperkalaemia and also that of thrombophlebitis and embolus.

2 Anorexia and occasionally vomiting occur.

3 Infrequently small bowel haemorrhage, ulceration (enteric coated) and stricture formation occur. All solid KCl preparations, with the possible exception of Diumide-K Continus, should be regarded as liable to cause gastrointestinal erosive mucosal lesions including the microencapsulated form. Only liquid KCl elixir is not associated with erosive lesions but is least well tolerated. The incidence of erosions is increased by drugs which delay gastric emptying. In general oral K should be taken with food.

4 Slow K has produced oesophageal ulceration.

If K is required as a supplement to diuretics then a K conserving diuretic may be more appropriate, for example spironolactone or amiloride.

WHO/UNICEF oral rehydration solutions have been mentioned already. There have been reports of hyperkalaemia following their use in children under one year of age. Hypernatraemia may also occur and it is recommended that these solutions are not used for longer than 24 hours without plasma electrolyte measurement.

Intravenous potassium

Again KCl is the drug of choice. Intravenous (i.v.) K is hazardous. Generally infusion rate should not exceed 10 mmol hour^{-1} or 120 mmol daily.

In cases of severe depletion maximum infusion of 30 mmol hour^{-1} for 1–2 hours only is permissible. K is usually added to an i.v. infusion in a concentration of 40 mmol l^{-1}. It should never be added to blood, blood products, mannitol or solutions of amino acids or lipids because it may cause lysis of red cells or may precipitate substances from solution.

If K is added to fluid in plastic containers it must be very well mixed otherwise it remains concentrated at the site of addition to the solution with potentially fatal consequences. Frequent blood samples should be taken for evaluation of K status and monitoring futher therapy. If ECG signs of T wave peaking appear then the infusion should be stopped or slowed.

Intraperitoneal administration

If peritoneal dialysis is being undertaken for acute renal failure it is likely that plasma K will be high and therefore the dialysis fluid should have a low K content. Both 61 and 62 dialysates contain 0.26 g KCl per litre. Peritoneal dialysis may rarely be used for severe refractory CCF to remove excess oedema fluid and then additional K may be added to the dialysate to maintain plasma K. This is of particular importance when the patient is taking digitalis and large doses of diuretics.

Other measures to reduce potassium loss

1 During treatment of hypertension small doses of diuretic should be used. Bendrofluazide 5 mg is often an adequate dose.
2 In mild heart failure a small dose of a loop diuretic produces less K loss than a thiazide diuretic for the same fluid loss. If larger doses are required volume depletion should be avoided since this will increase aldosterone production with further K loss.
3 A moderate reduction in salt intake (70–80 mmol daily) reduces K loss by two mechanisms:
(a) It reduces the dose of diuretic required.
(b) K secretion in the distal renal tubule depends on the amount of Na delivered to this site.

Potassium chloride is one of the most frequently prescribed drugs in medical practice but should be used with great care. An American physician in 1973 argued that more lives were lost by potassium therapy than saved by it. After the controversy regarding K replacement with diuretics in the treatment of low normal K came the question of whether K supplements reduce hypertension. This is discussed in Chapter 7 but in summary it seems that in some studies of hypertensive patients K supplements do reduce BP. High K intake reduces Na reabsorption by the tubule and in hypertensive patients so treated may be associated with weight loss due to reduction in ECF volume and total body Na. K suppresses renin and is a vasodilator by virtue of a direct effect on smooth muscle of arterioles.

It seems logical to correct diuretic induced hypokalaemia in hypertensive patients and if hypokalaemia aggravates hypertension perhaps diuretics should NOT be used routinely as first line therapy. There is wide variation in response to K supplements and doubt that diuretic induced hypokalaemia can easily be corrected by K supplements. The dangers of hyperkalaemia are particularly important if non-prescription K supplements are used, for example, as an alternative to table salt.

In Bartter's syndrome which is hyperaldosteronism, hypokalaemia, alkalosis, normotension, juxtaglomerular apparatus hyperplasia and increased renin, captopril has proved successful in treatment. Excess renal prostaglandin may be the underlying abnormality. There is associated angiotensin resistance, impaired tubular NaCl reabsorption and low urinary Ca excretion.

In hypokalaemic patients a deficit of magnesium is a very likely clinical accompaniment.

HYPERKALAEMIA

This is defined as that K concentration which exceeds the upper limit of the normal range for that reporting laboratory which effectively is usually greater than $5.5 \, \text{mmol} \, \text{l}^{-1}$. Acute hyperkalaemia is a medical emergency requiring prompt recognition and treatment. Therefore it is important to rule out artifacts producing a falsely raised plasma K.

Causes of hyperkalaemia

These fall into four groups although often several predisposing causes are present together:
1 Spurious.
2 Increased intake.
3 Reduced excretion.
4 Redistribution of K, release from cells.

Spurious

Haemolysis.
Thrombocytosis.
Massive leukocytosis.
Muscle exercise during venous occlusion.
The hyperkalaemia of haematological disorders is now being
referred to as pseudo hyperkalaemia.

Increased intake

Iatrogenic, very rapid, i.v. load. ′
Excessive oral intake.
Several drugs such as penicillin salts and proprietary cough
medicines contain large amounts of K but rarely produce severe
toxicity. Salt substitutes for patients requiring low salt diets often
contain considerable quantities of K. This may be of significance in
cases of mild to moderate renal failure. The remarks on oral
rehydrating solutions are relevant here. Oral Ca supplements may
also contain a significant amount of K, for example 4.5 mmol in
each Sandocal tablet.

Reduced excretion

This is by far the most important cause of hyperkalaemia occuring
in:
Acute renal failure.
Severe chronic renal failure.
Sodium depletion.
Steroid deficiency:
    Addison's disease
    hypoaldosteronism.
Drugs — Cyclosporin, non-steroidal anti-inflammatory drugs.
Inappropriate use of K conserving diuretics such as Triamterene,
spironolactone, amiloride.
The kidney has a very large reserve for K excretion and hyper-
kalaemia is uncommon until over 90% of renal function is lost and
GFR is less than 20 ml min^{-1}. Both Addison's disease and iso-
lated hypoaldosteronism can present with severe hyperkalaemia.

In the latter condition there is a normal glucocorticoid response to ACTH. Non-steroidal anti-inflammatory drugs especially indomethacin, are associated with hyperkalaemia especially if used in the elderly population with mild pre-existing uraemia.

Redistribution of K, release from cells

Acidosis, diabetic ketoacidosis.
Muscle injury, catabolism.
Suxamethonium.
Digitalis.
Beta-blocking drugs.
Leukaemia chemotherapy.
Hyperkalaemic periodic paralysis.
Pathological haemolysis:
  incompatible blood transfusion
  autoimmune states
  disseminated intravascular coagulation
  malaria (falciparum)
  sodium chlorate poisoning.
Malignant hyperthermia.

Plasma K rises during muscle exercise and falls fairly rapidly after it stops. Transient hyperkalaemia is not uncommon after marathon running. Beta adrenoreceptors moderate the acute hyperkalaemia of exercise whereas alpha adrenergic receptors enhance hyperkalaemia and may protect against hypokalaemia when exercise ceases. After beta-blockade with propranolol the rise in K is greater. Whether exercise in patients on propranolol whose K increases to 6 mmol l^{-1} or more is harmful is not certain. The onset of muscle fatigue may be protective.

Acidosis is at its worst following cardiac arrest. The influence of acid base status on K balance is complex, related mainly to changes in acidity of total body fluids but also to changes in anion composition of the plasma, osmolality and pancreatic and adrenal hormones.

When extensive muscle damage occurs for example in crush injury then plasma K may rise much faster than excretion or redistribution can cope with the rise. The depolarizing muscle relaxant drug suxamethonium produces fasciculation of muscle prior to paralysis. This results in an abrupt rise in plasma K especially in patients with catabolic illness, burns, muscle trauma, spinal cord injury and renal failure. Prior administration of a nondepolarizing muscle relaxant such as curare, can obtund this response. Insulin dependent diabetics can develop very severe hyperkalaemia if insulin is withheld perioperatively. Hypertonic agents such as urea or saline can increase K.

Hypothermic patients often exhibit hypokalaemia due to shift of K from ECF to ICF. K therapy in hypothermia should only replace measured losses otherwise profound hyperkalaemia may occur on rewarming with cardiac dysrhythmia.

Continuous recording of plasma K with an intravascular electrode shows an acute transient increase in K with administration of $CaCl_2$ and as this drug may be used in resuscitation of an acidotic patient it is a potentially dangerous phenomenon.

Excessive cell breakdown due to any cause will produce hyperkalaemia but K stores in the body remain normal or low.

There is a condition of benign hereditary hyperkalaemia which is inherited as an autosomal dominant characteristic.

Signs and symptoms of hyperkalaemia

Most patients are asymptomatic until there is a marked rise in plasma K (greater than 6 mmol l^{-1}) when the following symptoms develop:

Muscle weakness, loss of tendon reflexes and rarely paralysis.

Listlessness.

Mental confusion.

Tingling, numbness and paraesthesia, particularly affecting the circumoral region, and the lower limbs.

Nausea, vomiting and occasionally ileus.

The ECG is the best indicator of hyperkalaemia.

ECG signs of hyperkalaemia

1 Tall tented T waves.
2 Wide QRS.
3 Widening PR.

Cardiac arrest may occur (in asystole) at plasma K levels greater than 7 mmol^{-1}.

Hyperkalaemia is never a diagnosis made clinically and it can be seen that many of the symptoms are identical with those of hypokalaemia. However, it does much more harm than hypokalaemia although children and those with chronic impairment of renal function tolerate hyperkalaemia better than previously normal adults.

Management

The first priority is to make the diagnosis by measurement of plasma K or by looking for characteristic ECG abnormalities. These, however, give no indication of whether the abnormality is one of excess total body K for example in renal failure or mal-distribution of K which may be associated with significant total body K depletion such as occurs in severe catabolism.

Immediate treatment

Severe hyperkalaemia requires correction within minutes. There-fore, measures to increase urinary excretion are inappropriate as they are too slow. Redistribution of K is the best way of dealing with hyperkalaemia together with counteraction of the cardiac effects with calcium. This should precede administration of $NaHCO_3$ which may produce a dramatic change in pH and precipi-tate hypocalcaemia.

1 Intravenous calcium chloride or gluconate (5 mmol) will antagonize the cardiotoxic effects of hyperkalaemia. This dose may need to be repeated and continuous ECG monitoring is mandatory.

2 50 g glucose i.v. with soluble insulin 25 units will increase K uptake into cells along with glucose facilitated by insulin. When insulin is used with glucose it is important to rule out Addison's disease beforehand as dangerous hypoglycaemia may occur.

3 $NaHCO_3$ 50 ml 8.4% may be given initially. This is hypertonic and should be given into a large vein. It corrects acidosis and returns K to the ICF although it must be remembered that a sodium load may be dangerous in renal failure.

Intermediate treatment

If renal failure is present any body fluid volume deficits should be corrected. Sepsis should be treated urgently and attempts made to provide adequate nutrition to reverse hypercatabolism.

Dialysis

Both haemodialysis and peritoneal dialysis will readily reduce plasma K and correct acidosis.

Ion exchange resins

Resonium A (sodium resin) or calcium resin are available. These may be given orally or rectally. Enemas should be retained for at least 30 minutes to permit exchange.

In digoxin toxicity, digoxin antibodies have successfully been used to reduce hyperkalaemia by reversal of inhibition of ATPase.

CHRONIC HYPERKALAEMIA

These situations are in the main related to renal failure where dialysis and transplantation are employed in management. Otherwise, restriction of dietary protein and K, prevention of acidosis and sepsis and correction of fluid depletion are all important.

Careless administration of K supplements and K conserving diuretics should be avoided. 9-fludrocortisone (a mineralocorticoid) corrects isolated hypoaldosteronism and Addison's disease responds to glucocorticoid replacement.

During acute treatment regular measurement of plasma K concentration and continuous ECG monitoring are mandatory as a swing to hypokalaemia is not uncommon. This situation used to be seen during treatment of diabetic ketoacidosis with large doses of insulin.

Potassium chloride is a dangerous solution. Accidental injection into the epidural space has been reported and resulted in severe pain and permanent paraplegia. An ampoule of KCl was also used accidentally to dissolve an antibiotic resulting in severe pain after 4 mmol had been administered. Concentrated KCl should be stored and used with great care and kept well away from more innocuous solutions.

VALUE OF URINE POTASSIUM MEASUREMENT

This may be helpful in assessment of patients with hypokalaemia and to evaluate the route of K loss. If hypokalaemia is due to K deficit then a urine K greater than 10 mmol l^{-1} suggests that the kidney is responsible for the loss unless there has been insufficient time for renal conservation to take place. Such a situation may occur in diuretic administration, metabolic acidosis or alkalosis, some renal tubular diseases, hyperadrenocorticism and occasionally in leukaemia and carbenicillin administration. If urine K is less than 10 mmol l^{-1} then the gastrointestinal tract is the likely source of K loss. Estimation of urine potassium will also be a guide to the severity of catabolism and the stress response to trauma. Postoperatively urine K losses may be very high. It is equally important, however, to measure other sources of K loss for example the K content of fistula losses as a guide to replacement, and always to bear in mind the effect of acid-base status on K balance.

Chapter 9
Abnormal Calcium, Phosphate and Magnesium Balance

CALCIUM

Abnormalities of Ca, PO_4 and Mg are closely linked. In essential hypertension, for example, plasma renin activity shows continuous negative correlation with serum Mg and positive correlation with serum ionized Ca. Renin may contribute to changes in Ca and Mg flux across cell membranes. The absolute value of plasma Ca varies with pH and albumin concentration; a low plasma albumin reduces the amount of bound Ca and hence results in a low plasma Ca (see Chapter 4).

Hypercalcaemia

Definition

Total plasma Ca greater than 2.55 mmol l^{-1}.

Causes

1 Malignancy.
2 Primary hyperparathyroidism.
3 Other causes:

vitamin D intoxication	sarcoidosis, probably due to extra-renal production of 1,25(OH)$_2$D$_3$
pulmonary tuberculosis	renal failure
thiazide diuretics	berylliosis
tertiary hyperparathyroidism	vitamin A intoxication
hyperthyroidism	idiopathic hypercalcaemia of infancy

immobilization   post-renal transplantation
acromegaly   familial hypocalciuric
                                            hypercalcaemia
phaeochromocytoma   complication of parenteral
                                            nutrition in renal failure
lithium therapy
Paget's disease
granulomata after cosmetic silicone injection

Addisonian crisis and the milk alkali syndrome are rare causes of hypercalcaemia as is familial hypocalciuric (benign) hypercalcaemia in which daily urinary Ca excretion is less than 5 mmol but which usually requires no treatment.

It is important to exclude false hypercalcaemia due to venous stasis, the nonfasting state and polycythaemia.

MALIGNANT HYPERCALCAEMIA

This is the commonest cause of hypercalcaemia forming 50% of all cases and is often accompanied by low serum albumin.

1 Many primary tumours metastasize to bone causing bone destruction and release of Ca into the plasma. Commonest of these are carcinoma of the breast and bronchus. Multiple myeloma produces a similar picture.

2 Tumours may release a hormone with parathormone like activity. The commonest tumours with these effects are carcinoma of the bronchus and hypernephroma.

The mechanism of hypercalcaemia may be:

(a) Mobilization of Ca from bone.

(b) Reduced loss of Ca by the kidney.

(c) Increased intestinal absorption of Ca.

Many malignant tumours do secrete factors which resorb bone by enhancement of osteoclastic activity. These include prostaglandins and interleukin I. In normal circumstances this would be dealt with by normal hormonal mechanisms. Thus there must be a substance which prevents renal clearance of Ca which is immunologically unlike parathyroid hormone but biologically has a PTH-like action on the kidney. True ectopic PTH secretion is extremely rare.

There is not usually increased intestinal absorption of Ca. Ectopic production of $1,25(OH)_2D_3$ by Hodgkins and other lymphomas also occurs. In general therefore the hypercalcaemia is due to the increased bone resorption which exceeds the capacity of the kidney to excrete it.

PRIMARY HYPERPARATHYROIDISM

30% of newly diagnosed cases of hypercalcaemia are due to primary hyperparathyroidism (HPT) and some 20% of these have renal calculi. 70% of cases are due to parathyroid adenoma, 19% to hyperplasia and 4% to carcinoma. Now that many patients have a plasma Ca measured on a multichannel analyser more cases of hypercalcaemia are being detected. Many of these cases have only a mild increase in Ca concentration. 57% of patients in a recent series presented as a chance finding. Most at risk appear to be females over 70 years of age. 14% presented with a hypercalcaemic syndrome of dehydration and confusion (see below). Treatment with lithium may be associated with HPT with or without an adenoma and familial HPT is inherited as an autosomal recessive characteristic. Normal elderly people tend to have a high $25(OH)D_3$ and low PTH concentration although in the sick patient this is reversed. HPT is a difficult diagnosis to make in this age group and high PTH concentration must be interpreted in the light of the patient's general health and nutritional status. Serum immunological PTH is the best discriminant in the differential diagnosis between raised Ca of malignancy and primary HPT. In some circumstances plasma Ca may be raised only intermittently. These can be identified by a Ca tolerance test.

Calcium tolerance test. 1 g of Ca is given orally. In HPT there is hyperabsorption of Ca, marked hypercalcaemia and hypercalciuria. In addition plasma $1,25(OH)_2D_3$ levels are raised and there is abnormal suppression of PTH.

The hydrocortisone suppression test suppresses the increase in Ca due to nonparathyroid malignant disease and HPT with bone disease but not that due to HPT without bone disease. A definitive diagnosis of HPT may be made from an increase in immunoreactive PTH in the plasma of a hypercalcaemic patient or on biochemical criteria alone.

Biochemical features of hyperparathyroidism
1 Raised plasma Ca.
2 Decreased renal tubular reabsorption of PO_4
3 Increased renal tubular reabsorption of Ca.
4 Increased Ca absorption from the intestine.
5 Raised plasma $1,25(OH)_2D_3$.
6 Increased urine hydroxyproline excretion.
7 Increased plasma levels of bone alkaline phosphatase.

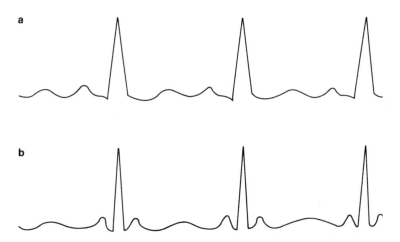

Fig. 9.1 ECG changes due to an abnormal plasma calcium. Moderate to severe hypercalcaemia shortens the QT interval on the ECG and increases susceptibility to digitalis induced arrythmia. Heart block occasionally develops. Severe hypercalcaemia leads to merging of the S and T waves and eventually cardiac arrest. (a), hypocalcaemia; (b), hypercalcaemia.

In familial hypocalciuric hypercalcaemia there is usually normal PTH, normal or increased Mg, normal or low PO_4 and low urinary excretion of Ca. In immobilized patients, urinary Ca excretion is markedly elevated and $1,25(OH)_2D_3$ strikingly reduced. This may be an example of resorptive calciuria.

Signs and symptoms of hypercalcaemia

Anorexia, nausea and vomiting.
Weight loss.
Peptic ulcer, abdominal pain.

Constipation.

Pancreatitis.

Renal concentrating defect.

Renal calculi.

Tiredness, muscle weakness, neuromuscular paralysis.

Cardiac dysrythmias, shortening of the QT interval seems relatively rare.

Ectopic calcification — cornea, conjuctiva, skin. Nephrocalcinosis and adrenal calcification occur.

Bone and joint pain.

Other symptoms of hypercalcaemia include headache, paraesthesia, apathy, mental disturbances and amenorrhoea. Vomiting further impairs electrolyte balance and worsens dehydration. The renal concentrating defect produces polyuria, polydipsia and dehydration.

The kidney can excrete large amounts of Ca but renal function becomes impaired because of the deleterious effects of hypercalcaemia on all parts of the nephron.

Effects of hypercalcaemia on the kidney

1   Reduced GFR due to renal vasoconstriction.

2   Salt loss due to inhibition of active reabsorption in the distal tubule.

3   Water loss due to antagonism of the action of ADH on the collecting duct.

4   Volume depletion increases proximal reabsorption of Na and Ca thereby perpetuating and worsening the existing hypercalcaemia.

Management of hypercalcaemia

Severe hypercalcaemia is a medical emergency associated with decreasing renal function. Treatment should be directed at the underlying cause but if severe hypercalcaemia is present urgent therapy should be started without awaiting the underlying diagnosis. When the balance between destruction and formation of bone is disturbed acutely the term 'disequilibrium hypercalcaemia' is used; the hypercalcaemia is unstable and may present as hypercalcaemic crisis.

CORRECTION OF DEHYDRATION AND OTHER ELECTROLYTE ABNORMALITIES

The patient may be very dehydrated at presentation but although large volumes of fluid may be required it is important to bear in mind that in elderly patients there is a risk of coexisting cardiac and renal disease and that cardiac failure may be precipitated by over-enthusiastic fluid loads.

FORCED SODIUM DIURESIS

This has been advocated to increase GFR and reduce renal tubular reabsorption of Ca. 4–6 litres of 0.9% sodium chloride are given daily with 20–40 mg i.v. frusemide every 4–8 hours. Frusemide further increases the loss of Ca. This technique requires very careful monitoring because of the increased urinary loss of salt and water. Urine electrolyte measurements are particularly important since large losses occur and plasma K and Mg may be reduced to dangerously low levels. In severe hypercalcaemia bone is the usual source of Ca and the next logical step is to suppress bone resorption with phosphate.

PHOSPHATE

This increases the movement of Ca into bone but some extraskeletal Ca deposits may also occur especially in overtreatment. The risks are less when phosphate is given orally, for example as effervescent tablets. A suitable dose is 0.5–1.5 g neutral phosphate daily. If the patient is in coma then 50 mmol phosphate may be given i.v. over 6–8 hours. Plasma Ca may begin to fall within minutes and the effect may last several hours after the end of the infusion. A maximum of 100 mmol daily should be given. Nausea and diarrhoea may occur.

Dangers
1 Metastatic calcification.
2 Hypocalcaemia.
3 Hypotension.
4 Oliguric renal failure.
These are reduced by limiting the plasma phosphate level to 2 mmol l^{-1}.

CALCITONIN

This suppresses bone formation with relatively little toxicity. Three forms are available:
1 Porcine.
2 Human.
3 Salmon.

Dose. 4 MRC units kg^{-1} intramuscularly 12 hourly or in an adult 100 MRC units 12 hourly. 20% of all cases fail to respond and reductions of more than 0.7 mmol l^{-1} are rare. Use of non-human calcitonin results in development of antibodies in 50% of patients which often causes resistance to treatment. Intranasal administration of calcitonin may be a useful alternative means of administration. Unwanted effects are common and include flushing, pain on injection, diarrhoea, vomiting and abdominal pain. Many of these symptoms suggest excess 5-hydroxytryptamine and do indeed respond to the 5-HT antagonist pizotifen.

STEROIDS

These reduce intestinal absorption of Ca. The hypercalcaemia of sarcoidosis which is associated with raised $1,25(OH)_2D_3$ and vitamin D intoxication responds as do some cases of malignant disease. Effects, however, may be slow in onset. Prostaglandin synthetase inhibitors such as flurbiprofen reduce Ca in sarcoidosis.

MITHRAMYCIN

Mithramycin should be used in a dose of 25 µg kg^{-1} intramuscularly or by i.v. infusion. This should be repeated after 24–48 hours once only if the first dose is ineffective. This form of treatment is often very effective *but* with continued treatment marrow depression, renal and hepatic damage may occur.

Intravenous chelating agents are not used nowadays.

DIPHOSPHONATES

Clodronate disodium, a new diphosphonate, reduces Ca in HPT with suppression of bone disease. In Paget's disease diphosphonates reduce excessive bone resorption with a reduction in pain and also reduce hypercalcaemia. Short term treatment with 20 mg kg^{-1} daily may maximize suppression of disease activity whilst reducing exposure to unwanted effects such as osteomalacia. Diphosphonates may give rise to renal impairment due to formation of aggregated complexes between cations and the drug.

DIALYSIS

This is a short term manoeuvre to buy time for other measures to take effect. It will also be useful if acute renal failure has already occurred or in the presence of CCF.

CIMETIDINE

This has recently been added to the treatment of HPT to reduce gastric acidity.

Hypercalcaemia of malignant disease is often difficult to treat; steroids and a high fluid intake are the mainstays of therapy although the response to steroids is slow. Use of calcitonin alone is disappointing although combined with steroids there is a rapid Ca lowering effect due to acute reduction in renal tubular absorption. Prostaglandin synthetase inhibitors have a rather unpredictable response and whereas mithramycin works well and reliably, long term treatment results in marrow depression. For long term management oral neutral phosphate can be used and the newer diphosphonates seem promising. Specific PTH antagonists are being developed to counteract the renal component responsible for the hypercalcaemia.

Where hypercalcaemia is due to vitamin D poisoning induction of hepatic enzymes may prove successful in management, using for example glutethimide.

Surgery

Surgery is indicated in HPT for:
1 Osteitis fibrosa.
2 Recurrent renal stones.
3 Hypertension, reduced renal function.
4 Peptic ulceration.

5 Pancreatitis.

6 Psychiatric disturbance.

Mild cases of HPT may not require treatment but there is currently no way of predicting which of them will develop hypertension or renal damage. It is said that hypercalcaemic patients identified by screening have a much lower prevalence of stones than those seen before screening. If such patients are hypertensive treatment with a beta adrenergic blocking drug may reduce both the blood pressure and the PTH level thus reducing hypercalcaemia. Thiazides, however, will increase plasma Ca with adverse consequences. The natural history of untreated HPT is more benign than originally thought but the best treatment of a fit patient with symptoms is surgery performed by an experienced parathyroid surgeon. This may also be the treatment of choice for a young (<60 years) symptomless patient. The elderly symptomless patients may be followed up only and drug therapy not used at all. A very few are now treated with oral PO_4 and oestrogens or clodronate used with caution.

Renal calculi

80% of renal calculi in Britain are made up of calcium oxalate or phosphate or both. Recurrent renal calculi are commonest with underlying conditions such as the following:

1 Hypercalciuria.

2 Medullary sponge kidney.

3 Renal tubular acidosis.

4 Primary hyperoxaluria.

Two forms of the condition previously called idiopathic hypercalciuria exist:

(a) Absorptive hypercalciuria.

(b) Renal hypercalciuria.

Excretion of Ca is >0.1 mmol kg^{-1} daily on a diet without excess salt or protein.

In absorptive hypercalciuria, where plasma Ca is normal and PO_4 low or normal, the primary abnormality seems to be increased intestinal absorption of Ca since parathyroid function is normal or suppressed and vitamin D is not concerned. In renal hypercalciuria, fasting hypercalciuria exists with parathyroid stimulation in the presence of normocalcaemia. PTH stimulates renal $1,25(OH)_2D_3$ production with increased intestinal absorption of Ca. It may be that the underlying defect is reduced renal tubular absorption of PO_4.

Considerable reduction in hypercalcaemia can be achieved by dietary restriction of Ca which also improves the efficacy of Ca binders. Urinary Ca excretion may be reduced by inorganic phosphate. Sodium cellulose phosphate will bind Ca in the intestine. Thiazide administration also reduces hypercalciuria. Treatment should be continued indefinitely. Therapy should also be directed at the other underlying causes of recurrent renal calculi.

Hypocalcaemia

Definition

This occurs when the total plasma Ca is less than 2.20 mmol l^{-1}.

Causes

1 Hypoparathyroidism. This may be primary, idiopathic or secondary to accidental removal of the parathyroid glands at thyroidectomy.

2 Pseudohypoparathyroidism. This is due to target organ resistance to normal levels of PTH.

3 Magnesium deficiency, usually following extensive resection of the intestine or prolonged parenteral nutrition without adequate supplements (see Chapter 12).

4 Vitamin D deficiency.

(a) Reduced oral intake or absorption; vitamin D_3 is formed by the action of sunlight on ergosterol in the skin and D_2 is found in foodstuffs such as oily fish, margarine, eggs and milk concentrates. To develop vitamin D deficiency therefore one has to be deficient in both dietary sources and sunlight. This tends to occur in Asian immigrants of all ages and in the elderly confined to home.

(b) Disturbed metabolism. The metabolism of vitamin D is discussed in Chapter 4. In advanced renal failure there is a reduction in $1,25(OH)_2D_3$ production and this is a major factor in the development of renal osteodystrophy. Some drugs such as phenytoin act as hepatic enzyme inducers and, by induction of the hepatic microsomol P450 system, speed the turnover of vitamin D, facilitating production of less active metabolites thereby reducing its effects and resulting in osteomalacia.

Glucagon by virtue of its stimulation of calcitonin induces hypocalcaemia. Glucocorticoids and malabsorption states are also associated with hypocalcaemia.

There is an association between vitamin D and haemopoiesis. Vitamin D deficiency is associated with anaemia, myelofibrosis and extramedullary haemopoiesis. These abnormalities are reversed by vitamin D therapy. $1,25(OH)_2D_3$ seems to have a role in differentiation of marrow precursor cells and an effect on monocytes which are precursors of osteoclasts.

5   Renal failure.

6   Other causes of hypocalcaemia include acute pancreatitis, sodium citrate or EDTA administration, osteoblastic secondary bone deposits and malignancy treated with cytotoxic agents. This may be accompanied by hypomagnesaemia and hypoparathyroidism. Infants receiving cows milk derivatives with a high phosphate content may become hypocalcaemic. Massive infections of subcutaneous tissues, burns during slough and early granulation and generalized peritonitis produce hypocalcaemia because large amounts of Ca are immobilized in the diseased tissues and exudates. In septic shock a lowered ionized Ca concentration occurs which is correlated with a reduction in cardiac output. Any surgical operation may produce transient hypocalcaemia. Hypoalbuminaemia reduces bound Ca only. Radioprotective and chemoprotective agents inhibit secretion of PTH, enhance calciuria and produce hypocalcaemia. Pernicious anaemia may also be accompanied by hypocalcaemia, although associated hypoparathyroidism must be excluded.

MYASTHENIC STATES ASSOCIATED WITH IONIC IMBALANCE

Most often these are due to acute shifts or losses of K but severe Ca depletion may produce similar acute muscle weakness because of the loss of the calcium facilitating effect on acetylcholine release. Administration of magnesium has a similar effect on neuromuscular transmission to that of Ca depletion. In certain critical situations such as cardiopulmonary bypass, continuous measurement of Ca ion concentration has been used. The technique is difficult owing to changes in pH and temperature which affect Ca-protein binding.

Signs and symptoms of hypocalcaemia

NEUROMUSCULAR FUNCTION

Neuromuscular function is severely impaired with paraesthesiae, muscle cramps and tetany.

TETANY

This is spasm of muscle usually due to low ionized Ca due to hypocalcaemia or severe alkalosis from overbreathing or the vomiting of pyloric stenosis. Latent tetany can be demonstrated by Trousseau's sign or Chvostek's test but if tetany is severe there may be spontaneous spasm of the muscles of the hand (main d'accoucheur) or carpopedal spasm. In infants laryngeal stridor occurs. Patients may present with abdominal pain due to spasm of these muscles.

OTHER FEATURES OF HYPOCALCAEMIA

Tingling, numbness and pins and needles in the hands and face are often worse after overnight fasting. Epilepsy may be the presenting feature especially in children and after prolonged lactation. Generalized electroencephalogram (EEG) changes may be present. Oculogyric crises and an extrapyramidal syndrome may occur.

It is classical that unless severe, hypocalcaemia is not diagnosed for years by which time personality changes with depression and irritability may be present.

Intracranial calcification, cataracts and systemic moniliasis occur in hypoparathyroidism.

Idiopathic hypoparathyroidism may be associated with Addison's disease. Rickets with knock knees, bow legs and rickety rosary occurs in children and osteomalacia in adults or children.

Associated phosphate abnormalities are described in the next section. Magnesium deficiency should always be suspected when hypocalcaemia is diagnosed.

ECG CHANGES OF HYPOCALCAEMIA

Hypocalcaemia delays ventrical repolarization and increases the QT interval and ST segment (see Fig. 9.1). Heart block and ventricular arrythmias may develop. The heart may be refractory to digoxin.

Treatment of hypocalcaemia

As far as possible this is treatment of the underlying cause.

CALCIUM

Oral supplements should be given in a dose of 22 mmol daily (1 g of elemental Ca). Even with oral therapy repeated plasma Ca levels should be measured.

In acute hypocalcaemia or in situations where myocardial function is compromised treatment with intravenous Ca salts may be required. Two preparations are available:
1 Calcium gluconate.
2 Calcium chloride.
Calcium chloride has advantages: the body's retention of this salt is greater and more predictable than calcium gluconate. The positive ionotropic effect of chloride is greater than gluconate which produces an unpredictable increase in ionized Ca. Both preparations are available in 10 ml ampoules containing 5–10 mmol of Ca ($CaCl_2$) or 2.23 mmol (calcium gluconate). Administration of i.v. Ca is *dangerous* and should only be done under ECG control *slowly*. Particular care should be taken in acidotic patients or those receiving digitalis. Severe dysrhythmias can occur. If i.v. Ca is to be given by continuous infusion then it must on no account be added to blood, blood products, lipid or aminoacid solutions. Up to 5 mmol 8 hourly may be given in this way. For supplementation during total parenteral nutrition about 8 mmol are required daily.

VITAMIN D

Very careful monitoring is essential to avoid overtreatment. In circumstances such as rickets where renal function is normal vitamin D_3, cholecalciferol, 3000 units (75 μg) daily or 1.25 mg twice weekly will suffice. Providing renal function is normal this will be converted into $1,25(OH)_2D_3$ by the kidney and it is this metabolite which is most important for healing of vitamin D deficient rickets.

In the 1960s the prevalence of Asian rickets rose in the UK due to inadequate solar ultraviolet radiation, dietary and clothing habits. This is now falling again due to vitamin D supplementation to children and young adults. Pregnant Asian women especially should receive such supplements to prevent osteomalacia and congenital rickets. However vitamin D deficiency rickets and secondary HPT may not respond to $1,25(OH)_2D_3$ unless the diet is supplemented with Ca. Rickets and osteomalacia can present as hyperphosphataemia.

In renal osteodystrophy which was found by early workers to be resistant to vitamin D there is decreased production of $1,25(OH)_2D_3$. In addition $24,25(OH)_2D_3$ may play a part in bone formation and the level of this is depressed earlier in renal failure than $1,25(OH)_2D_3$. If this is the case both of these vitamin D metabolites will be necessary to prevent and heal renal osteodystrophy. Hypophosphataemic vitamin D resistant rickets may need PO_4 supplements and is associated with hypercalciuria. The hypophosphataemia may increase $1,25(OH)_2D_3$ which then increases Ca absorption, parathyroid suppression and hypercalciuria. This may be one end of the spectrum of hereditary absorptive hypercalciuria.

For hypoparathyroidism supplies of synthetic parathormone are now available but a vitamin D preparation remains the drug of choice. In the past 1–2 mg (40 000–80 000 units) vitamin D daily have been used. This therapy has a slow onset and is accompanied by hypercalcaemia which is sometimes prolonged. Dihydrotachysterol 0.25 mg daily and more recently $1,25(OH)_2D_3$ has been found to control plasma Ca rapidly. Again hypercalcaemia may be a problem but it responds within days to stopping the drug. Response to treatment should be monitored by serum and urine Ca and creatinine.

Coexisting magnesium and potassium depletion should be corrected.

Calcium and the heart

Ca levels may be very low following massive blood transfusion as the Ca is chelated by citrate. Usually Ca is rapidly restored from the large Ca pool in the skeleton but if the blood pressure is low and perfusion poor restoration may take some time and myocardial depression occur unless intravenous Ca is given.

In the resting state cell membranes are more permeable to K than to Na ions. Depolarization in a cardiac cell is associated with a rapid influx of Na. At a certain level of depolarization a slow flow of Ca ions prolongs depolarization to a plateau phase. Then the membrane becomes more permeable to K and less so to Na, repolarization occurs and the action potential declines.

Excitation of muscle is associated with release of Ca from stores and binding of Ca to the inhibitory troponin-tropomysin complex thence antagonizing its inhibitory action and muscle contraction occurs. If ECF Ca is prevented from entering the cell muscle contraction is prevented. In animals high concentrations of isoprenaline cause myocardial necrosis due to a flood of Ca ions into cells with excessive activation of Ca dependent intracellular ATPases, energy depletion, mitochondrial damage and finally cell necrosis. This is prevented by slow channel Ca blocking drugs, for example verapamil, and beta-blocking drugs. Hypocalcaemia can precipitate congestive cardiac failure in patients with pre-existing arterial disease. In hypertensive patients, serum ionized Ca has been found to be lower than in normotensives although others have found a positive correlation between Ca and BP.

Calcium antagonists

Drugs which prevent passage of Ca through channels in cell membranes will have two effects:
1   They may relax muscle and produce vasodilation.
2   They may alter cardiac rhythm (effect on depolarization).
Drugs which produce an effect by Ca antagonism:
    Verapamil.
    Nifedipine.
    Prenylamine.
    Perhexiline.
These drugs inhibit the interaction of Ca with binding proteins such as calmodulin. They may be expected to affect glycogen metabolism, haemostasis and they interact with neuromuscular blocking drugs to reduce requirements.

Verapamil blocks Ca flux in the atrioventricular node and is useful for treatment of supraventricular arrythmias. The ECG should be monitored continuously during intravenous administration. Subsequent administration of Ca causes reappearance of the dysrhythmia. The inotropic drug amrinone enhances Ca liberation into the sarcoplasm and Ca antagonists such as verapamil and nifedipine attenuate its action. Nifedipine is used in the treatment of angina and hypertension. Prenylamine is used for angina but in the presence of hypokalaemia is prone to cause arrythmias. There is some recent evidence that these drugs may be of value in exercise induced asthma. This may be due to Ca dependent mediator release. Ca antagonists act by blocking Ca channels and hence reducing post exercise bronchoconstriction.

The abnormalities of cation handling in hypertension are discussed in Chapter 7. It seems logical to antagonize the potential constrictor effect of increased intracellular Ca in the treatment of hypertension. The BP lowering effect of Ca antagonists is greater in hypertensive than in normotensive patients. If Ca antagonists are to be used increasingly for hypertension their independent action on other tissues should be taken into account.

BONE DISEASE

In metabolic bone disease with excessive bone resorption and inappropriate calcification more normal ossification can be achieved with diphosphonates or calcitonin.

Paget's disease. Calcitonin used to be the drug of choice in Paget's disease, but reduction in bone pain and return of bone turnover to normal values can be achieved with the newer diphosphonates which also reduce hypercalcaemia. A combination of the two drugs has also been used.

Osteoporosis. In osteoporosis about 30% of bone substance must be lost before a radiological diagnosis can be made although measurements of bone mass by total body Ca is a possibility. Such a technique also shows the heavy bone loss due to corticosteroids in conditions such as rheumatoid arthritis. Although lack of calcitonin may be an important factor in the rapid development of bone loss after the menopause and prophylactic treatment with calcitonin may prevent this there is no evidence that such treatment can reverse established osteoporosis. Calcitonin reduces the rate of skeletal absorption. Other measures which are of value in the postmenopausal patients are oestrogens, high dose Ca (over 2000 mg daily), fluoride and anabolic steroids. Combined oestrogen and progesterone are effective. Cigarette smoking increases hepatic metabolism of oestrogens which may contribute to the reported increased risk of osteoporosis amongst smokers. For prevention of osteoporosis most attention should be focused on Ca supplements. Ca and fluoride can restore mineralization in severe osteoporosis or those who have already sustained fractures. Fluoride stimulates osteoblasts and positive Ca balance. In osteoporosis of bed rest, Ca, PO_4, diphosphonates and thiazide diuretics have all been tried and may reduce hypercalciuria without reducing bone loss although thiazide diuretics increase the mineral content of bone. Exercise is helpful and fluoride or oestrogens probably the least harmful. Diphosphonates may become the treatment of choice after more evaluation.

Vitamin D is not needed unless there is proven osteomalacia. There may be a case for fortifying food with Ca.

Pregnancy. If pregnancy occurs in very deprived conditions of grossly deficient Ca diets the mother's skeleton is sacrificed for the fetus and calcitonin levels are depressed. Normal pregnancy is associated with an increase in $1,25(OH)_2D_3$ which enhances Ca absorption from the intestine and raised calcitonin which reduces bone resorption.

Osteomalacia. Focal osteomalacia may occur as a side effect of Ca lowering treatment. It occurs after gastrectomy and is due to low vitamin D. Treatment is with vitamin D supplements and occasionally Ca.

Immobilization and space flight lead to bone resorption although this usually results in only a small loss from the total body Ca pool.

Intranasal salmon calcitonin is now available and in a dose of 100–200 units increases excretion of Na, K, Ca and Cl. A second potent Ca lowering hormone is described which may also be useful in the treatment of bone disease.

CALCITONIN

Calcitonin may be raised in physiological circumstances such as pregnancy and childhood. It is classically elevated in medullary carcinoma of the thyroid but also in malignant disease outside the thyroid especially the lung. Patients in renal failure, those with pancreatitis, pernicious anaemia, Zollinger Ellison syndrome, pancreatitis and hyperglycaemia also have raised calcitonin concentrations.

Calmodulin has been discussed in Chapter 4. It plays an important role in secretion of fluid and electrolyte into the gut. Drugs such as loperamide inhibit activity of calmodulin activated phosphodiesterase. Opioids do not share this effect.

PHOSPHATE

Phosphorus is ubiquitous in human metabolism and yet its disorders are infrequent and closely related to those of Ca metabolism. Because it is so widely distributed within cells dietary deficiency with enteral nutrition is almost unknown.

Hyperphosphataemia

Definition

This exists when plasma phosphate (PO_4) is greater than 1.35 mmol l^{-1}.

Causes

1 Hypoparathyroidism.
(a) Primary, idiopathic.
(b) Secondary, postsurgical.
2 Pseudohypoparathyroidism.
3 Vitamin D toxicity.
4 Renal failure. At low GFR phosphate is retained and the resulting hyperphosphataemia may itself further impair renal function. In chronic renal failure (GFR less than 20 ml min^{-1}) the hyperphosphataemia may cause renal calcification by:
(a) Increasing the plasma Ca $\times$ PO_4 product.
(b) Causing hyperparathyroidism which would suppress renal tubular function and increase acidosis. Phosphate restriction is then appropriate.

5   Acromegaly.
6   High phosphate intakes.
7   Malignancy associated with cytotoxics.
8   Severe catabolic states.
9   Acidosis.
Hyperphosphataemia is more common in infants, children and postmenopausal women. A raised plasma PO_4 causes little in the way of symptoms.

Clinical manifestations

1   Ectopic calcification.
(a)   Corneal, conjunctival.
(b)   Nephrocalcinosis.
(c)   Arterial calcification.
(d)   Skin calcification.
2   Secondary hyperparathyroidism may occur in renal failure.
3   Renal osteodystrophy. Specific therapy will be required in this situation.

Treatment

Treatment is that of the underlying cause and much of it appears in the section above on Ca.

Measures to reduce PO_4 in renal failure

1   Aluminium hydroxide. This binds PO_4 in the gastrointestinal tract thereby reducing absorption. It must be remembered that hypophosphataemia is also a danger and that aluminium is toxic to the central nervous system in chronic renal failure producing an encephalopathy known as dialysis dementia.
2   Renal dialysis will reduce PO_4 levels.

Relation between phosphate and acidosis

In acidosis H ion concentration is increased. This suppresses glycolysis by inhibition of phosphofrucktokinase. Less adenosine triphosphate (ATP) is synthesized, less inorganic phosphate is utilized and more is therefore available in the cells and plasma.

The increase in H within the cell is initially balanced by a high PO_4. Then the H is buffered by intracellular protein and PO_4 moves out of the cell to maintain the Gibbs Donnan equilibrium. The increased PO_4 outside the cell is excreted via the kidney. Acidosis also promotes PO_4 loss by exchange of H for sodium via disodium hydrogen phosphate in the renal tubules.

A rather similar situation occurs in respiratory failure with a low arterial oxygen. In these circumstances aerobic metabolism is reduced, ATP synthesis is reduced and again plasma PO_4 rises and is presented to the kidney for excretion. When the oxygen dissociation curve is shifted to the right as occurs in acidosis there is an increase in deoxygenated haemoglobin which is a very efficient buffer and enhances PO_4 loss from cells.

Hypophosphataemia

This exists when plasma PO_4 is less than 0.8 mmol l^{-1}. This situation is of much greater importance clinically than an excess of this anion because of the importance of PO_4 within the cell. It occurs in over 2% of general hospital patients.

Causes

1 Primary hyperparathyroidism and ectopic parathormone production.
2 Vitamin D deficiency.
3 Renal tubular defects; congenital and acquired.
4 Hyperventilation.
5 Alkalosis.
6 Low phosphate intake.
7 Glucose and insulin.
8 Adrenaline.
9 Glucocorticoids.
10 Chronic alcoholism.
11 Haemodialysis.
12 Postoperative situations.
13 Treatment of diabetic ketoacidosis with bicarbonate.
14 Septicaemia.
15 Beri-beri.
These causes may be grouped into:

(a) Reduced intake.
(b) Increased loss.
(c) Intercompartmental shift.

Reduced PO_4 intake is unlikely under normal circumstances but is very important in a patient receiving parenteral nutrition with inadequate supplements. Absorption of PO_4 can be reduced by magnesium or aluminium salts. Increased loss via the renal tubules occurs in HPT, steroid therapy and renal tubular insufficiency. Acidosis has already been discussed. Intercompartmental shifts occur when PO_4 enters the cell with glucose during a glucose infusion. It is then utilized in glycolysis and therefore the available phosphate is reduced. Hence hypophosphataemia may present a problem in postoperative surgical patients. The hypophosphataemia is maximal on the second or third postoperative day. The subsequent reduction in 2,3-diphosphoglycerate (2,3-DPG) is accompanied by increased haemoglobin affinity for oxygen and possible tissue hypoxia. This hypophosphataemia may be due in part to a change in the maximal tubular resorption for phosphate. Enzyme induction can alter metabolism of vitamin D and this mechanism is thought to be the cause of hypophosphataemia with certain drugs including mianserin.

Clinical and biochemical features of hypophosphataemia

Symptoms are unusual until plasma phosphate falls below 0.32 mmol l^{-1}.

Paraesthesiae, muscle weakness.
Rickets, osteomalacia.
Impaired intermediary metabolism:
  defective glycogenolysis
  defective glycolysis
  reduced intracellular ATP.
Increased purine degradation.
Rigidity of erythrocyte membranes.
Reduced erythrocyte 2,3 DPG, impaired oxygen release.
Impaired polymorphonuclear leucocyte function, reduced chemotactic, phagocytic and bactericidal activity.
Reduced renal tubular bicarbonate reabsorption.
Seizures.
Coma.

In the acutely ill patient in an intensive care unit hypophosphataemia is especially likely. If such a patient fails to breath adequately following a period of mechanical ventilation then measurement of plasma PO_4 should be undertaken without delay. If acute hypophosphataemia is superimposed on muscle dysfunction then potentially reversible muscle cell damage may be converted to irreversible necrosis.

Acute severe hypophosphataemia can mimic Wernicke's encephalopathy and can account for absent central nervous system function. It is essential therefore to exclude a metabolic cause such as this before proceeding to a diagnosis of brainstem death.

Management of hypophosphataemia

This hinges upon recognition of circumstances in which it is likely to occur and confirmation of the diagnosis. It is very important in this respect to ensure that blood for plasma PO_4 estimation is NOT taken when a glucose infusion is in progress.

PHOSPHATE ADMINISTRATION

PO_4 may be given orally or intravenously.

1 Oral phosphate may be administered as 0.5–1.5 g inorganic PO_4 daily.

2 Since many situations are associated with potassium depletion the appropriate intravenous replacement is K_2HPO_4. This contains 20 mmol K and 10 mmol PO_4 in 10 ml. Another solution available for use in this country is a 500 ml polyfusor containing:

Na   162 mmol l^{-1}
K   19 mmol l^{-1}
PO_4 100 mmol l^{-1}

This solution, however, provides rather a large sodium load. Daily requirements of PO_4 are 0.5–0.75 mmol kg^{-1}. In the acute situation 30–50 mmol daily should be given. To replace losses in addition to daily requirements 50–100 mmol may be required over 6–12 hours and this dose should be repeated after 24 hours. Some authorities recommend as much as 50 mmol for each 1000 calories during parenteral nutrition suggesting that overspill into the urine denotes adequate replacement. Over-enthusiastic replacement should be avoided as hyperphosphataemia may result in widespread metastatic calcification and severe hypocalcaemia. Intravenous PO_4 should not be given to hypocalcaemic patients.

MAGNESIUM

Hypermagnesaemia

Hypermagnesaemia exists when plasma Mg is greater than 0.9 mmol l^{-1}. This is rare and occurs mainly in chronic renal failure especially if magnesium containing antacids have been used inadvertently. It may also occur during acute renal failure although during the diuretic phase Mg levels may fall below normal. Mg should be administered with caution in the presence of latent renal disease especially if repeated administration is necessary.

Causes

1   Acute and chronic renal failure.
2   Uncontrolled diabetes mellitus.
3   Adrenocortical insufficiency.
4   Metabolic acidosis.
5   Mg containing medications for example antacids, cathartics, treatment for eclampsia.
6   Spurious hypermagnesaemia due to the sample being taken with excess venous stasis.

Clinical features of hypermagnesaemia

Impaired neuromuscular activity:
   muscle weakness
   loss of deep tendon reflexes.
   impaired autonomic nerve transmission.
   increase in the excitatory threshold.
   may be potentiated by aminoglycoside antibiotics.
Vasodilatation leading to hypotension.
Drowsiness.
ECG changes.
Mg levels greater than 2.5 mmol l^{-1} impair atrioventricular and intraventricular conduction. ECG change include an increase

in PR interval, broadening of the QRS complex and an increase in the height of the T wave. Mg levels greater than 5 mmol l^{-1} produce muscle paralysis similar to that due to curare. The amount of acetylcholine liberated from the neuromuscular junction is reduced, the sensitivity of the end plate to acetylcholine is impaired with decreased excitability of the muscle membrane. This action is due to competition with Ca for binding. It may be that the action of Mg in prevention of premature labour is due to displacement of Ca from binding sites and prevention of its action.

Respiratory centre depression also occurs in hypermagnesaemia. Mg levels greater than 8 mmol l^{-1} are associated with coma and cardiac arrest. Cardiopulmonary arrest is reported due to acute maternal hypermagnesaemia when $MgSO_4$ was used in the treatment of pre-eclampsia. The hypocalcaemia associated with hypermagnesaemia may be due in part to the suppressive effects of high Mg on PTH secretion.

Treatment of hypermagnesaemia

When cardiac conduction defects occur urgent treatment is required.
1 In renal failure dialysis is the treatment of choice.
2 In an emergency situation with dysrhythmia i.v. Ca may be given cautiously under ECG control in a dose of 2.5–5 mmol.
3 An enforced Na/Ca diuresis can be attempted along the lines of that described for hypercalcaemia but this must be undertaken with great care and frequent electrolyte analysis, especially if cardiac or renal function are impaired.

REGIME FOR FORCED DIURESIS

4–6 l 0.9% sodium chloride daily.
20–40 mg frusemide i.v. 4–8 hourly.
2.5–5 mmol Ca should be given every 4–8 hours in a litre of 0.9% saline.

Hypomagnesaemia

Definition

Hypomagnesaemia exists when plasma Mg is less than 0.7 mmol l^{-1}.

Causes

1 *Reduced oral intake.*
   Prolonged parenteral nutrition with inadequate supplements.
   Profound disturbances of oral intake such as anorexia nervosa.
2 Gastrointestinal disturbances.
   Vomiting.
   Diarrhoea.
   Prolonged nasogastric suction.
   Malabsorption — Mg soap formation will reduce absorption as
      will administration of large amounts of Ca.
   Fistulae.
   Protein calorie malnutrition.
   Alcoholic cirrhosis.
   Pancreatitis.
3 Endocrine.
   Hyperparathyroidism, hypoparathyroidism.
   Hyperaldosteronism — primary and secondary.
   Hyperthyroidism.
   Diabetic coma.
   Inappropriate or excess ADH.
   Disorders of lactation.
4 Renal disease.
   Glomerulonephritis.
   Pyelonephritis.
   Hydronephrosis.
   Renal tubular acidosis.
5 Alcoholism.
6 Other causes include osmotic diuresis, diuretic therapy, malig-
nant osteolytic disease, malnutrition and porphyria. Hypo-
magnesaemia may be idiopathic, familial or occur in the newborn.
Spurious hypomagnesaemia occurs in haemodilution and severe
hypoalbuminaemia.
   Alchohol affects the tubular mechanisms essential for reabsorption of Mg and increases production of metabolic intermediates
such as lactate which bind Mg. In diabetic coma Mg shows the
same changes as K. It is raised initially and falls during treatment
to potentially very low levels. Diuretic induced Mg deficiency may
also render the kidney unresponsive to diuretics. Captopril overcomes this to some extent and is also Mg sparing. K sparing
diuretics also spare Mg loss and may even produce Mg retention.

Clinical features

Symptoms are erratic and no well defined clinical picture exists but the following may be found with a plasma level of 0.3–0.5 mmol l^{-1} or less:

1 Nausea, vomiting and abdominal pain.
2 Neuromuscular excitability.
(a) Tetany, cramps
(b) Paraesthesia.
(c) Reversible neuropathy in dialysed patients
3 Muscle weakness with a coarse irregular tremor.
4 Lethargy.
5 Central nervous system effects include depression, irritability, ataxia and vertigo. Personality change with anxiety may develop and later confusion, hallucinations, coma and convulsions. Reversible neuropathy occurs in dialysed patients. Fitting may be the presenting feature especially in alcoholics.
6 Tachycardia may occur with either hypertension or hypotension. Cardiac dysrythmias such as ventricular ectopics are not uncommon. ECG changes include ST segment depression, T wave inversion and a prolonged QT interval. There is exacerbation of digoxin toxicity. A poor cardiac output state may occur which responds to Mg infusion. Mg deficiency may increase free intracellular Ca concentration, thus increasing vascular tone and reactivity. Some studies have shown a negative correlation between mean BP and serum Mg. Patients on long term diuretics receiving Mg supplements showed a significant fall in BP which may be a direct effect or via internal balance of Na, K, or Ca. Mg deficiency in soft water areas is believed to increase the number of deaths from ischaemic heart disease, although trials do not confirm that oral Mg leads to a fall in BP.

Mg is required for normal activity of Na/K ATPase and in hypomagnesaemia there will be partial depolarization of cells and increased irritability in excitable tissues. This also limits the amount of oral or i.v. K getting into cells in hypokalaemia and perpetuates the consequences of K deficiency such as heart failure and resistant dysrhythmia. Mg must therefore be corrected as well as K otherwise a misleading rise in plasma K will occur without correction of total body K.

Hypomagnesaemia impairs PTH function by reducing hormone

secretion with a blunted response to hypocalcaemia and resistance to the peripheral action of PTH. Blood levels of PTH rise rapidly after i.v. Mg administration.

Tetany due to low Mg may be more likely than hypocalcaemic tetany to produce convulsions. Administration of Ca in hypomagnesaemic tetany will not improve the situation until Mg concentrations are restored. Gentamicin causes hypomagnesaemia, hypocalcaemia and hypokalaemia as do some other antibiotics. Many clinical features listed here are in fact common to a number of electrolyte abnormalities and both hypokalaemia and hypocalcaemia may be associated abnormalities.

Cyclosporin neurotoxicity is associated with hypomagnesaemia. Adequate Mg levels are necessary for glucose disposal in response to insulin and hypomagnesaemia may account for the insulin resistance in management of diabetic ketoacidosis.

The development of symptoms depends on the plasma concentration and the ratio of this to other ions present. Mg deficiency is manifested more readily after correction of abnormalities of Na, K, Cl, Ca and acid base balance.

During total parenteral nutrition, as the patient becomes anabolic and Mg requirements increase, the deficiency is made worse. Mg like K is an intracellular ion so that the plasma levels poorly reflect the total body situation and considerable deficiency can occur with a normal plasma Mg. The Mg content of red blood cells correlates better with clinical signs of deficiency but is still unreliable. 24 hour urine excretion of Mg may be very low in deficiency states (0.5 mmol). Normally most of an administered dose of Mg is excreted but in deficiency states 40–80% is retained.

Assessment of magnesium depletion

Failure to excrete more than 70% of an administered load of 20 mmol Mg within 16 hours indicates depletion. Alternatively 0.25 mmol kg^{-1} may be given by infusion over 1–2 hours. If less than 80% is excreted within 24 hours Mg depletion exists.

Treatment of hypomagnesaemia

Magnesium sulphate is available in 10, 20 or 50% solution. Oral administration may produce diarrhoea although up to 70 mmol daily can usually be given safely by this route. Daily administration of $0.1–1.0$ mmol kg^{-1} is a suitable dose although correction of the intracellular deficit may take 1–2 weeks. Intravenous administration is essential to supply supplements during parenteral nutrition and may also be indicated in the treatment of malignant hyperthermia, digoxin intoxication and eclampsia. During parenteral nutrition 0.04 mmol kg^{-1} are required daily but up to 1.0 mmol kg^{-1} can be safely given daily and in an emergency this dose can be given over 4 hours. Oral supplements are usually adequate if hypomagnesaemia is due to excessive diuretic therapy.

It has recently been shown that in hyponatraemic patients with severe CCF on long term diuretic therapy, infusion of Mg results in an increase in plasma Na level, and an increase in muscle K. This is probably mediated through an effect of Mg on membrane ATPase.

Chapter 10
Disturbances of Acid Base Balance

In the management of patients with severe acid base disturbance it is essential to keep in mind the clinical condition of the patient when evaluating laboratory results. An isolated pH measurement is totally valueless without an arterial HCO_3 and in the case of respiratory disease a $PaCO_2$ as well. Free H ions are constantly being produced by the body but changes in H ion concentration are kept to a minimum by buffering mechanisms as already discussed. Respiratory and renal compensatory mechanisms come into play when buffering capacity is exceeded in an attempt to achieve a normal plasma pH.

Classically disturbances of acid base balance are divided into respiratory or metabolic acidosis or alkalosis and may be considered in relation to the Henderson–Hasselbalch equation:

$$pH = pK + \log \frac{HCO}{H_2CO_3} \qquad (1)$$

Normal plasma HCO_3 is 24 mmol l^{-1} and H_2CO_3 is 1.2 mmol l^{-1} so that

$$\log \frac{HCO}{H_2CO_3} = \frac{\log 24}{1.2}$$

$$= \log 20 = 1.301$$

therefore pH = 6.1 + 1.301 = 7.40.

METABOLIC (NON-RESPIRATORY) ACIDOSIS

This occurs when an abnormal amount of *non*-carbonic acid is formed or an abnormal loss of base occurs. Typical values are:

pH less than 7.36 (H ion concentration more than 44 mmol l^{-1})

$PaCO_2$ less than 4.6 kPa (35 mmHg)
HCO_3 less than 18 mmol l^{-1}

An increase in anion gap may occur (see below). In metabolic acidosis the value of

$$\log \frac{HCO_3}{H_2CO_3}$$

will be reduced and hence pH falls.

Causes of metabolic acidosis

Accumulation of acid

1 Diabetic ketoacidosis when blood levels of acetoacetate, acetone and beta-hydroxybutyrate are increased.
2 Lactic acidosis.
3 Renal failure when sulphate and phosphate radicles accumulate.
4 Salicylate overdose.
5 Following cardiac arrest. Anaerobic metabolism occurs in hypoxic situations and in conjunction with impaired tissue perfusion H ions are generated.

Substances such as paraldehyde, methyl alcohol, ethylene glycol, fructose, sorbitol, formalin, xylitol and ethanol may cause metabolic acidosis. Cyanide has a high affinity for the ferric ion of cytochrome c3 in mitochondria, inhibiting electron transfer and therefore oxidative phosphorylation and oxygen utilization. This results in a profound lactic acidosis due to anaerobic respiration often with accompanying hyperglycaemia. There may also be an increase in H ion concentration from accumulated hydrochloric acid released during the metabolism of arginine and lysine in synthetic amino acid solutions and following the therapeutic use of ammonium chloride. D-lactic acidosis can occur in short bowel syndrome and responds to antibiotic administration which suggests that it is secondary to bacterial production of d-lactate by abnormal gut flora; probably species of streptococcus and lactobacillus.

Loss of bicarbonate

1 From the gastrointestinal tract:
(a)   Small intestinal, pancreatic or biliary fistulae.
(b)   Diarrhoea.
(c)   Cholestyramine.
2 From the kidney:
(a)   Renal tubular acidosis.
(b)   Carbonic anhydrase inhibitors (Acetazolamide).
Carbonic anhydrase inhibitors are often used in the elderly for treatment of glaucoma but will also inhibit carbonic anhydrase in the renal tubule and result in a bicarbonate diuresis. They should be used with caution in the elderly with renal impairment because of the risk of acidosis. Carbonic anhydrase inhibitors in combination with salicylates in normal doses can induce metabolic acidosis in young people with previously normal renal function.

The body will try to compensate for this fall in pH by stimulation of respiration and a fall in $PaCO_2$ which will return the pH towards normal by reduction in H_2CO_3. This is termed respiratory compensation. Although the fall in pH is modified it does not return completely to normal.

Clinical effects of acidosis

1 Decreased cardiac output.
2 Pulmonary hypertension.
3 Cardiac dysrhythmias and arrest.
4 Oliguria.
5 Increase in circulating catecholamines.
6 Mental changes.
Experimentally induced metabolic acidosis due to infusion of NH_4Cl produces the following disturbances of renal electrolyte transport.
1 Inhibition of Na reabsorption in the distal renal tubule independent of aldosterone or filtered Cl load or volume expansion. Thus Na excretion is increased without any change in GFR, renin or aldosterone and not due to volume expansion.
2 Even in the presence of a rise in plasma K the excretion of K is inhibited.

3 Tubular Ca reabsorption is inhibited. This effect is independent of changes in parathormone secretion.

Treatment of metabolic acidosis

Treatment is directed at the underlying cause, and much of this has already been discussed in other chapters. Therapy with $NaHCO_3$ should be reserved for fairly severe disorders since it is not without hazard.

Sodium bicarbonate ($NaHCO_3$)

$NaHCO_3$ solution is available in a variety of concentrations. That commonly used for resuscitation is 8.4% which contains 1 mmol ml^{-1} of Na and HCO_3 and therefore there is a danger of the high Na load precipitating heart failure. The solution is hypertonic and irritant to veins resulting in extensive skin necrosis if the solution leaks from the vein into the tissues. A hyperosmolality syndrome may be precipitated, therefore 8.4% $NaHCO_3$ should be reserved for acute situations such as cardiac arrest, should preferably be given via a central line and in circumstances in which fluid restriction is appropriate.

Several formulae are available for calculation of the appropriate dose of bicarbonate; one uses the base deficit $\times$ one third body weight. It is prudent to partially correct deficits and repeat blood gas analysis. An alternative formula is $(1/2CO_2 - HCO_3) \times 1/2$ body weight.

OTHER DISADVANTAGES OF NAHCO3

1 In uncontrolled diabetes mellitus, acidosis exists with a low level of 2,3 DPG. Acidosis shifts the oxygen dissociation curve to the right and a low 2,3 DPG shifts it to the left so that the effect of these two abnormalities cancel each other out and oxygen delivery and tissue oxygenation remain normal. If $NaHCO_3$ is administered in these circumstances the pH will rise leaving the unopposed effect of a low 2,3 DPG to shift the oxygen dissociation curve to the left with impaired delivery of oxygen to the tissues. 2,3 DPG levels take several days to return to normal and it is important in this respect to ensure a normal plasma phosphate.

2   In diabetes $NaHCO_3$ may precipitate the disequilibrium syndrome. This is discussed in Chapter 14.

3   Rebound alkalosis may occur if excessive doses of $NaHCO_3$ are used.

4   Administration of $NaHCO_3$ results in a rise in $PaCO_2$. If the patient is able to hyperventilate the excess CO_2 can be excreted via the lungs. In patients with an impaired conscious level mechanical ventilation at a high minute volume may be required with repeated measurement of blood gas tensions.

5   In a hypokalaemic patient $NHCO_3$ administration will increase pH, promote further K uptake into the cell and lethal hypokalaemia may occur.

General recommendations for bicarbonate therapy

First complete evaluation, both clinical and biochemical, must be undertaken.

After confirming the diagnosis, alkali therapy is not used until pH is <7.2.

Electrolyte balance, hydration and osmolality should be evaluated concurrently.

Continuous cautious therapy should be closely monitored.

In summary, $NaHCO_3$ is a dangerous drug but may be essential after cardiac arrest for the successful action of inotropic agents.

There is a possibility that administration of base by the CSF route may ameliorate cellular acidosis due to ischaemia and improve survival of cells.

Trihydroxymethylaminomethane (THAM)

THAM is a non-sodium containing buffer which has been used in some parts of the world with a variable degree of success to treat metabolic acidosis.

Other supportive therapy is important such as treatment of cyanide poisoning, hyperventilation or dialysis.

LACTIC ACIDOSIS

Production and utilization of lactate involve conversion to pyruvate. There are methods of rapid bedside diagnosis of lactate in plasma or whole blood.

Lactic acidosis may be broadly classified into type A where there is inadequate delivery of oxygen to the tissues and lactate is generated faster than it can be removed or type B where tissue hypoxia is not relevant.

Causes of type A lactic acidosis

1   Exercise.
2   Shock.
3   Hypoxia (PaO_2 less than 35 mmHg).
4   Anaemia.
5   Congestive cardiac failure.

Causes of type B lactic acidosis

1   Drugs: phenformin, ethanol, etomidate, outdated tetracycline, probably due to degradation products and paracetamol poisoning.
2   Intravenous feeding using excessive doses of sorbitol or fructose.
3   Diabetes.
4   Renal failure.
5   Liver disease.
6   Infection especially septicaemia.
7   Leukaemia, lymphoma.
8   Thiamine deficiency.
9   Hereditary: glucose-6-phosphate deficiency.

The commonest cause of acute serious type B lactic acidosis is biguanide therapy for diabetes. Phenformin is ten times more likely than metformin to produce an attack of lactic acidosis (0.64 cases per 1000 patient years of treatment for phenformin compared to 0.05–0.08 for metformin). The mortality of type B acidosis is 50%.

Whenever possible an attack should be prevented. This involves abandoning phenformin in favour of metformin and using glucose rather than fructose or sorbitol as the energy substrate for parenteral nutrition. The danger with fructose infusion increases at an infusion rate greater than $0.5 \text{ g kg}^{-1} \text{ hour}^{-1}$.

Treatment of lactic acidosis should be directed at the underlying cause ensuring adequate oxygen delivery to the tissues at all times. Alkalinization with $NaHCO_3$ is the mainstay of therapy. Isotonic $NaHCO_3$ (1.4%) should be used to bring the pH back to normal over about 6 hours. Often over 1000 mmol bicarbonate may be required. If hyperkalaemia coexists this therapy will be beneficial as the K will enter the cells as the pH rises. If hyperglycaemia exists insulin may be required. Insulin with glucose is potentially useful in treatment of phenformin induced lactic acidosis.

Dichloroacetate reduces lactate in humans with an increase in bicarbonate, arterial pH and improvement in blood pressure and cardiac output. This is a useful adjunct to therapy but prognosis depends upon the underlying disease. Dichloroacetate acts by stimulating activity of pyruvate dehydrogenase, the enzyme that catalyses the rate limiting step in oxidation of lactate and pyruvate and can be tried when lactate exceeds 5 mmol l^{-1}. Bicarbonate therapy may be without effect or deleterious. Lactate behaves like any other weak acid and alkali therapy favours a shift of lactate from ICF to ECF so that an increase in lactate after bicarbonate therapy does not necessarily indicate a worsening metabolic picture. Other alternative therapies include mechanical hyperventilation to increase pH above a critical value below which myocardial depression and dysrhythmias occur. The extent of lactic acidosis is a good prognostic indicator of severity of the underlying disease.

A central venous pressure line and urinary catheter should be inserted to monitor progress as circulatory overload with cardiac failure is a serious complication. Haemodialysis may be required to treat cardiac failure. Bicarbonate dialysis and fluid removal are ideal. Repeated estimations of blood gases are essential in any disturbance of acid base balance and in these circumstances are of greater practical value than repeated lactate levels.

METABOLIC (NON-RESPIRATORY) ALKALOSIS

This occurs due to excess production of base or abnormal loss of non-carbonic acid. Typical findings are:
    pH greater than 7.44

$PaCO_2$ greater than 6.0 kPa (45 mmHg)

H ion concentration less than 36 mmol l^{-1}

HCO_3 ion concentration greater than 32 mmol l^{-1}

In this situation the third component in the Henderson–Hasselbalch equation (1) is increased, therefore log HCO_3^- is increased and pH rises. Compensation occurs by hypoventilation so the H_2CO_3 also rises thereby modifying the increase in pH.

Causes of metabolic alkalosis

1 Loss of H ions.

Renal

  Primary and secondary hyperaldosteronism and K depletion

  Conn's syndrome

  Cushing's syndrome

  Bartter's syndrome

  ACTH secreting tumour.

Drugs

  diuretics — thiazides, frusemide, ethacrynic acid

  corticosteroids

  carbenoxolone.

Liquorice.

Post hypercapnoea.

Gastrointestinal

  Nasogastric suction

  Vomiting

  High intestinal obstruction.

  Villous adenoma.

  Congenital.

2 Gain in alkali.

Iatrogenic milk alkali syndrome

Administration of $NaHCO_3$.

Metabolic conversion of organic acid anions to HCO_3, e.g. lactate and citrate.

This situation is seen after a large blood transfusion when the citrate is metabolized over the next 48 hours to HCO_3 resulting in metabolic alkalosis.

The bicarbonate concentration of certain medications is high, especially fizzy soluble preparations such as panadol soluble which contains 18 mmol per tablet or Hedex with 14–18 mmol in each tablet. Patients with renal impairment are particularly at risk.

A metabolic alkalosis is perpetuated when there is a reduction in ECF volume, where excess mineralocorticoid activity occurs with K depletion especially when this becomes severe (greater than 450 mmol total body K deficit). In these situations renal excretion of HCO_3 is reduced. Metabolic alkalosis is almost always accompanied by a low chloride, which maintains electrical neutrality in the presence of a raised HCO_3.

Urine chloride

This will be low when hydrochloric acid is lost from the stomach, when intravenous $NaHCO_3$ or diuretics have been administered but high in hyperaldosteronism, Bartter's syndrome, Cushing's syndrome, liquorice ingestion and severe prolonged K deficiency. These disturbances are further discussed elsewhere.

Clinical effects of alkalosis

Tetany, Trousseau and Chvostek signs are demonstrable at pH > 7.55
Hypocapnic vasoconstriction.
Left shift in oxygen dissociation curve.
Mental changes.
Hypokalaemia.

Kidney

In the proximal tubule there is obligatory Na reabsorption the extent of which is controlled by ECF volume. In the distal tubule Na is reabsorbed in exchange for K or H ion and influenced by aldosterone, such that if hypokalaemia occurs there will be increased H ion excretion. When renal retention of HCO_3 occurs correction of the pH alone without treatment of the underlying disease will result in recurrence and persistence of the metabolic alkalosis.

Treatment of metabolic alkalosis

Severe alkalosis may be life threatening especially if it is accompanied by hypokalaemia, but it is uncommon. Treatment includes the following:

1 Restoration of ECF volume. This may involve transfusion of NaCl, plasma or blood. It is important to give chloride. NaCl being simplest if a Na load is not contraindicated.

2 Restoration of plasma and whole body K concentration — using KCl or K conserving diuretics such as triamterene or amiloride if indicated.

3 Inhibition of aldosterone where appropriate, using spironolactone.

4 Inhibition of carbonic anhydrase by acetazolamide which will produce retention of H ions.

5 Direct acidification. HCl, NH_4Cl, lysine or arginine hydrochloride may be used after the other measures listed above have been completed. These may all be given intravenously and will result in the release of free H ions. HCl should only be administered through a central venous line at a rate of 0.2 mmol H kg^{-1} $hour^{-1}$ until pH is less than 7.45. A maximum dose of 300–350 mmol day^{-1} should not be exceeded. A significant reduction in $PaCO_2$ and oxygen consumption occurs due to improved cellular metabolism.

It is important to remember that the compensatory increase in $PaCO_2$ which occurs in metabolic alkalosis may be life saving and must be allowed to fall gradually during treatment. Under no circumstances should the patient be mechanically ventilated.

HCl has also been used in hypercapnia with compensatory HCO_3 retention with claims for improved oxygenation and ventilation. The outstanding danger is reduction in pH due to metabolic acidosis superadded to the existing respiratory acidosis. There may be difficulties using amino-acid solutions and perhaps fat emulsions during treatment of metabolic alkalosis.

In animals with metabolic alkalosis the jejunum shows increased net absorption of K, whereas the ileum reduces net Na and K and increases HCO_3 absorption. The opposite effects occur in acidosis. In the colon in alkalosis there is net secretion of HCO_3 with increased chloride absorption and reduced K secretion.

Intracellular carbonic anhydrase activity and concentrations of H, HCO_3 and Ca are mediated by calmodulin. Systemic pH changes may alter the effects of neural and humoral mediators of intestinal transport.

RESPIRATORY ACIDOSIS

$$pH = pK + \log \frac{HCO^-_3}{H_2CO_3} \qquad (1)$$

This is due to an excess of CO_2 and may be acute or chronic; in the latter situation renal compensation occurs. CO_2 becomes hydrated to H_2CO_3 so that the value log HCO_3^- falls with a fall in pH. The kidney compensates by retaining HCO_3 so reducing the fall in pH.

Causes of respiratory acidosis

1 Chronic obstructive pulmonary disease (COPD).
2 Respiratory centre depression, for example, narcotic analgesic overdose.
3 Impaired neuromuscular function:
Neuromuscular blocking drugs, myasthenia gravis.
4 Excess CO_2 in inspired gas mixtures. This is only likely to occur during anaesthesia with malfunctioning apparatus.
5 Increased CO_2 production — fever, oxidation of the concentrated dextrose solutions used for energy substrates during parenteral nutrition.
6 Restrictive chest wall or lung disease.
7 Adult respiratory distress syndrome in its later stages. Typical blood gases in the acute situation of respiratory depression are as follows:

pH	7.2
$PaCO_2$	9.46 kPa
	(70 mmHg)
PO_2	9.46 kPa
	(70 mmHg)
HCO_3	29 mmol l^{-1}

In the more common COPD where renal compensation has occurred typical blood gases are:

pH   7.30
$PaCO_2$ 8.0 kPa
        (60 mmHg)
PO_2 9.46 kPa
        (70 mmHg)
HCO_3 35 mmol l^{-1}

Treatment is aimed at the underlying cause for example improving ventilation using intermittent positive pressure ventilation if necessary. Alkali therapy has no place in chronic respiratory acidosis. In COPD the aim should be an H ion activity less than 56 mmol l^{-1} (pH $>$ 7.25) rather than a normal $PaCO_2$. It is the acidity of the blood which is important and this will have been modified by renal compensatory HCO_3 retention. Repeated blood gas monitoring is of course mandatory.

If intermittent positive pressure ventilation is used to produce a normal $PaCO_2$ the patient will be left with a raised HCO_3 and metabolic alkalosis.

RESPIRATORY ALKALOSIS

$$pH = pK + Log \frac{HCO_3^-}{H_2CO_3} \tag{1}$$

In this situation $PaCO_2$ is reduced so that

$$\log \frac{HCO_3^-}{H_2CO_3}$$

is increased with an increase in pH.

Causes of respiratory alkalosis

1 Hysterical hyperventilation.
2 Excessive intermittent hyperventilation.
3 Central nervous system disorder:
   head injury
   encephalitis

4 Tissue hypoxia, for example; anaemia, gram negative sepsis.
5 Interstitial pulmonary disease.
6 Pulmonary oedema.
7 Hepatic failure.
8 Salicylate intoxication

Pregnancy and acute mountain sickness also stimulate respiration. Compensatory changes occur if the disturbance is prolonged. The kidney excretes more HCO_3 to restore pH to normal by reducing

$$\log \frac{HCO_3^-}{H_2CO_3}$$

Again, treatment is that of the underlying cause, although mountain sickness may be prevented by acetazolamide.

In chronic alkalosis a rise in 2,3 DPG concentration prevents the increase in oxygen affinity which would occur due to a shift to the left of the oxygen dissociation curve because of the increase in pH.

Despite these apparent clear cut disturbances of acid base in practice there is often a mixed disorder, hence the importance of a good history and clinical examination. When confronted with an estimation of blood gases as with the plasma urea and electrolyte result it is important to isolate the most abnormal parameter since other abnormalities are possibly secondary to this or of a compensatory nature. In these circumstances an acid base diagram may be very helpful.

The Flenley acid base diagram (Fig. 10.1) is an alternative way of looking at acid base disturbances and is in fact a modification of the Siggaard Andersen nomogram in which the bicarbonate buffer system only is considered. The linear relationship between H ion activity and $PaCO_2$ is plotted on a graph in which isopleths of equal HCO_3 concentration radiate out as a fan from the origin. 95% confidence limits are shown on the diagram. Serial values of blood gases are plotted to define the nature of the disturbance and its progress.

On the acid base diagram a point above the normal pH of 7.4 implies an acidosis and a point below an alkalosis. The plasma bicarbonate can be read directly. The diagram is of great value during therapy to monitor the changes in response to treatment.

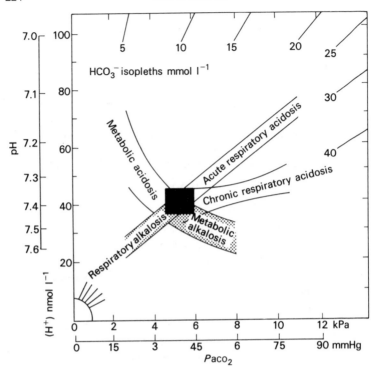

Fig. 10.1 Flenley acid base diagram.

ANION GAP

For electrical neutrality to exist the number of anions (negatively charged ions such as chloride and bicarbonate) must be equal to the number of cations (positively charged ions such as sodium and magnesium). The anion gap may be calculated in several different ways and the 'normal' range will therefore vary between laboratories. One method of calculation which can be done on a routine urea and electrolyte result is $(Na + K) - (Cl + HCO_3)$. This gives a value between 11–19 mmol l^{-1}. This value represents approximately the sum of unmeasured anions which are protein, phosphate, lactate and 3-hydroxybutyrate.

Some laboratories exclude K in calculation of the anion gap which gives a lower normal range.

Causes of an increased anion gap

1 Uraemic acidosis.
2 Ketoacidosis.
3 Salicylate poisoning.
4 Lactic acidosis.
5 Methanol, ethylene glycol and paraldehyde toxicity.
All of these causes are situations of acidosis but there are many other causes of acidosis in which the anion gap is not abnormal because chloride replaces bicarbonate, for example in diarrhoea.
6 Dehydration.
7 Therapy with sodium salts of strong acids where the acid anion is only slowly metabolized, for example sodium lactate. Antibiotics such as carbenicillin are administered as the sodium salt and the carbenicillin anion increases the anion gap. An anion gap greater than 30 mEq l^{-1} is usually due to an identifiable organic acidosis (lactic or keto-acidotic) although in 30% a more unusual cause requires detailed investigation.

Causes of low anion gap

1 Dilutional states.
2 Hypoalbuminaemia. Albumin at normal blood pH has a marked negative charge and therefore accounts for most of the anion gap.
3 Hypernatraemia, hypermagnesaemia, hypercalcaemia.
4 Paraproteinaemia. Here the increased viscosity of blood interferes with blood sampling.
5 Bromism. Some autoanalysers cannot distinguish between chloride and bromide so that an increase in bromide will present as a fall in the anion gap.

Minor variations in the anion gap should be interpreted with care since it is calculated from four variables. However, it is of value in detecting an abnormality before more specific investigation can be undertaken. Therefore the practice of discontinuing chloride measurement in routine U and E results is to be deprecated.

ASSESSMENT OF BLOOD GAS MEASUREMENTS

Study the following blood gas measurements and assess which respiratory or metabolic disorder of acid base status the information suggests. Answers are given at the foot of the page.

1 pH   7.27
   $PaCO_2$ 8.5 kPa (64 mmHg)
   HCO_3 37 mmol l^{-1}
   PaO_2 8.5 kPa (64 mmHg)
2 pH   7.50
   $PaCO_2$ 6 kPa (45 mmHg)
   HCO_3 40 mmol l^{-1}
   PaO_2 11 kPa (8.3 mmHg)
3 pH   7.55
   $PaCO_2$ 2.7 kPa (20 mmHg)
   HCO_3 22 mmol l^{-1}
4 pH   7.31
   $PaCO_2$ 4.0 kPa (30 mmHg)
   HCO_3 16 mmol l^{-1}
   $PaCO_2$ 13 kPa (98 mmHg)

(1) Respiratory acidosis.   (2) Metabolic alkalosis.
(3) Respiratory alkalosis.   (4) Metabolic acidosis.

Chapter 11
Fluid Balance in
the Surgical Patient

PREOPERATIVE ASSESSMENT, FLUID BALANCE AND RESUSCITATION

A patient presenting for minor or moderate elective surgery is unlikely to require preoperative fluid therapy. A patient who is already ill may have a variety of fluid and electrolyte disturbances.
1 Loss or gain of water causing disturbed osmolality.
2 Abnormal ECF volume, intravascular volume or cardiac output.
3 Disturbances in Na, Ca, K which are discussed elsewhere.

Assessment of the disorder

History

A history of fluid therapy and loss is invaluable. In the case of pyloric stenosis with vomiting, for example, predictable water, Na, K, Cl and H losses may occur and marked metabolic alkalosis with tetany may be present. Although an estimate of the loss of electrolyte may be made from known concentrations in different secretions wide variation does occur and measurement of the electrolyte content of losses in excess of 100 ml is essential.

FLUID BALANCE CHART

An accurate fluid balance chart is of great value in assessment of fluid requirements but except in special units it is also very difficult to find. Allowance must be made for insensible losses.

Clinical examination

A search should be made for dry mucous membranes or oedema. The pulse volume and blood pressure should be estimated and the

Table 11.1 Composition of secretions.

Secretion	Na	K mmol 1^{-1}	Cl	HCO$_3$	Volume (litres 24 hour $^{-1}$)
Parotid saliva	112	19	40	–	1.5
Gastric juice	50	15	140	0–15	2–3
Pancreatic juice	130	5	55	110	0.5–1
Bile	145	5	100	38	0.5–1
Ileal juice	140	11	70	Variable	–
Normal stool	20–40	30–60	20	–	0.1
Diarrhoea	30–140	30–70	–	20–80	Variable
Insensible sweat	12	10	12	–	0.5

If visible sweating occurs there is a marked increase in its sodium and chloride concentration.

chest examined for basal crepitations. Severe hidden fluid loss may occur in paralytic ileus.

Weight

Changes in body weight of 0.5–1 kg in 24 hours are usually due to loss or gain of water. Although it is a simple matter to weigh a healthy patient daily weighing of a sick patient is surprisingly difficult. Table 11.2 relates the degree of dehydration to weight loss.

Table 11.2 Assessment of fluid loss.

	Fluid loss in children (% body weight)	Fluid loss in adults (% body weight)
Mild dehydration	5	4
Moderate dehydration	10	6
Severe dehydration	15	8

This water loss is accompanied by some salt loss and is not usually associated with hypernatraemia. ECF depletion is the major result. All common fluid losses are roughly isotonic so strictly speaking the term 'dehydration' implying water loss only is inappropriate.

Biochemical measurement

Blood urea, haemoglobin and haematocrit will rise in dehydration. When this becomes severe plasma Na will also begin to rise. In conditions of plasma loss, for example pancreatitis, a haematocrit of 60% or over constitutes a life threatening emergency.

Urine output

A urine output of 0.5–1 ml kg^{-1} $hour^{-1}$ is acceptable. When an adult becomes oliguric 25 ml or less urine will be produced hourly. It is then essential to know whether this is due to hypovolaemia (reduced renal perfusion), to renal disease or to post renal obstructive factors.

OLIGURIA

Various investigations will help to elucidate the underlying cause.
1   Urine specific gravity >1.016 suggests a prerenal cause of oliguria.
2   Urine Na >30 mmol l^{-1} suggests intrinsic renal failure.
3   Blood urea >35 mmol l^{-1} suggests intrinsic renal failure.
4   Urine urea <185 mmol l^{-1} strongly suggests intrinsic renal failure.
5   Urine: plasma urea ratio. If oliguria has a prerenal cause this ratio will be >20:1.
(a)   In early intrinsic renal failure it will be <14:1.
(b)   In late intrinsic renal failure it will be <5:1.
6   Urine: plasma osmolality ratio. If oliguria has a prerenal cause this will be >2:1.
(a)   In early intrinsic renal failure it will be <1.7:1.
(b)   In late intrinsic renal failure it will be <1.1:1.

RESPONSE TO DIURETICS

The ability of the kidney to respond to a diuretic may be tested. 20 g mannitol (100 ml 20%) is infused over 20 minutes. The absence of a diuresis implies intrinsic renal failure. In a patient

who is already hypervolaemic mannitol administration may precipitate pulmonary oedema especially if it cannot be excreted. If
the central venous pressure (CVP) is very high 1 g of frusemide
should be given slowly instead.

It is unfortunate that we can only easily make measurements of
intravascular volume (CVP, urine output) which do not help in
making an assessment of the interstitial or intracellular space.

Central venous pressure

This is the pressure measured from the right atrium (RA) and it is
a reflection of the volume of the circulation and the ability of the
right heart myocardium to deal with the venous return. A low
CVP indicates hypovolaemia either absolute or relative, for example profound vasodilatation of histamine release. A high CVP
suggests failure of the right heart for whatever reason. Unfortunately the CVP may be misleading in hypovolaemic states due to the
intense vasoconstriction of both arterioles and capacitance vessels
which increases venous return and maintains the CVP within the
normal range (3–7 cm water above the right atrium) (Fig. 11.1).

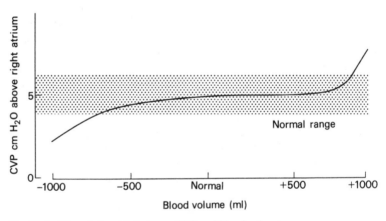

Fig. 11.1   The relationship between CVP and blood volume.

However, serial measurements of CVP are very valuable if
severe loss or gain of fluid is anticipated or in elderly patients with
cardiac or renal disease.

Unfortunately CVP measured in the RA does not always reflect the state of the left heart. A flotation catheter (Swan Ganz) may be necessary to measure pulmonary capillary wedge pressure (PCWP) which reflects left atrial pressure (LAP) and is normally 6–12 mmHg. This is the effective filling pressure or preload of the left side of the heart. When myocardial function is impaired measurement of LAP is a more useful and safer guide to blood volume replacement than CVP.

Very large volumes of fluid can be infused without a rise in CVP but with a marked increase in lung water and Na which can be very deleterious. Then a rise in LAP warns against further fluid administration and suggests that pulmonary oedema may occur. If the colloid osmotic pressure (COP) is normal pulmonary oedema occurs with a LAP of 20 mmHg or more.

Chest X-ray

The chest X-ray may show pulmonary oedema either interstitial or intra-alveolar.

Fluid losses

Normal preoperative 'nil by mouth' regimes do not usually cause significant water loss. Insensible fluid loss continues unabated and may be raised by pyrexia or high environmental temperature. Further loss may be incurred by diuretics, enemas, or vomiting due to premedication. Preoperative bowel preparation with mannitol can produce hypovolaemia and hypoglycaemia. Any fluid deficit should be corrected after the use of osmotic purgatives. Loss or gain of H_2O results in disturbed osmolality which can be corrected as already described in Chapter 6. Losses of isotonic saline, plasma proteins or blood cause reduction of intravascular volume and minimal change in tonicity. Such losses require urgent replacement preoperatively.

Intravascular volume, cardiac output and shock

A patient who has a reduced blood volume requires this to be replaced with the correct fluid prior to surgery, so as to maintain a stable cardiovascular system (CVS) throughout the procedure.

Shock

Shock exists when the cardiac output is inadequate to supply the nutritional demands to all tissues. In practice this involves inadequate blood flow to the tissues and results in cell damage leading to cardiac, pulmonary, metabolic, renal and microcirculatory complications. It does not necessarily mean that cardiac output is below normal.

Causes of shock

Hypovolaemic shock (heart normal)
   haemorrhage
   plasma loss, burns
   pancreatitis
   peritonitis.
Relative hypovolaemia
   septicaemia
   anaphylaxis.
Cardiogenic shock
   inadequate performance of the left ventricle (LV).

Clinical signs of hypovolaemic shock include a reduction in blood pressure (BP) and urine output, a rise in pulse rate and cold cyanosed extremities with slow capillary refill after blanching. In hypovolaemia renal blood flow shifts from the outer renal cortex to the inner juxtamedullary portions of the kidney where there are fewer glomeruli with longer loops of Henle. This shift in blood flow results in increased reabsorption of Na and water. As less filtrate is presented to the tubules the reabsorption of Na is further increased. Reduced Na concentration in the distal tubule and a reduction in perfusion pressure in the renal arterioles both stimulate the juxtaglomerular apparatus to secrete renin. This then increases aldosterone production. Hypovolaemia or impaired perfusion also stimulates ADH release.

MANAGEMENT

Management of hypovolaemic shock includes establishing the diagnosis followed by mechanical haemostasis and treatment of the underlying condition.

With CVP and urine output monitoring fluid is infused if the CVP is low (normal CVP is 3–7 cm water above RA) bearing in mind that the change in CVP with blood volume replacement is not linear (see Fig. 11.1). If there is doubt about the adequacy of the circulating blood volume 200 ml fluid can be infused rapidly (fluid loading test). If the CVP rises by more than 2 cm water for more than 10 minutes then no further fluid is required for the time being. Correlation between CVP and wedge pressure during fluid loading in septicaemic and hypovolaemic shock is only fair. If the CVP is greater than normal in shock and PCWP measured by Swan Ganz catheter is less than 12 mmHg then volume expansion is indicated. If PCWP and therefore left ventricular filling pressure is greater than 15 mmHg vasoactive or inotropic drugs are indicated. The desirable aim of treatment of circulatory failure is an increase in cardiac output. Drugs such as chlorpromazine may increase cardiac and urine output by afterload reduction and perhaps an effect on ADH. Phentolamine also reduces myocardial oxygen demand and may prevent the suppression of insulin release. Further discussion of the many vasoactive drugs of potential value in this situation is beyond the scope of this book.

The most important aspects of volume replacement are:
1   Early recognition of its necessity.
2   Rapid institution.
3   Frequent reassessment of the clinical state of the patient and revision of treatment.

FLUID AVAILABLE TO TREAT HYPOVOLAEMIC SHOCK

1   Blood.
2   Plasma.
3   Dextran.
4   Gelatin.
5   Hetastarch.
6   Crystalloid.

BLOOD

Blood of the correct group should be used to treat haemorrhagic shock if the haematocrit is less than 25% and should be continued until a satisfactory clinical state is achieved. The haematocrit of stored whole blood is 35% and that of packed cells is 70%. Stored blood is very unlike the patient's own circulating blood. It may be stored in a citrate anticoagulant with dextrose. Two such solutions were available:

1   Acid citrate dextrose (ACD) solution 120 ml. When this is mixed with 420 ml of blood the mixture has an initial pH of 7.4 at 4 °C but this falls during storage.

2   Citrate phosphate dextrose (CPD). This has certain advantages over ACD for blood preservation. These include slightly less haemolysis, less K leak from the cells and better post-transfusion survival of red blood cells. Twenty-day-old blood stored in CPD can be expected to have a pH of 6.71. This is a higher pH than in ACD blood and correlates with a long shelf life. Storage of blood reduces the number of red cells which survive after transfusion. 99% of red blood cells survive 24 hours post-transfusion if the blood is given immediately after collection, but only 75–80% survive 24 hours if the blood has been stored for 21 days.

2,3 Diphosphoglycerate (2,3 DPG). This is better maintained in CPD blood than in ACD blood mainly because of the greater pH of the CPD solution. Blood can be transfused up to 4 weeks after collection if stored in CPD. 2,3 DPG is important in affecting the position of the oxygen dissociation curve. A low 2,3 DPG level shifts the curve to the left with impaired release of oxygen in the tissues. The concentration of 2,3 DPG falls to zero in both ACD and CPD blood. With a small transfusion of old blood with its low level of 2,3 DPG there may be little clinical effect but if any deficiency exists in the oxygen delivery system for example a low haematocrit, reduced cardiac output or pulmonary disease then 2,3 DPG depletion may become clinically significant. The position of the oxygen dissociation curve also depends on Na and K. P_{50} correlates with

$$\frac{[Na] + [K]}{Hb}$$

Permeability of cells to K is inversely related to activity of 2,3 DPG phosphatase. Ionic permeability of erythrocytes is dependent on [H] partly through an effect of pH directly on the membrane and partly through an effect on glycolysis and ATP production.

SAG-M. A new storage medium for blood contains saline, adenine, glucose and mannitol. Red blood cells are preserved for up to 28 days at a PCV of 65–70%. SAG-M blood was originally intended for use with human albumin solution but the latter is currently in very short supply in this country.

Blood is stored at 4–6 °C and therefore requires warming prior to transfusion. It contains no ionized Ca although Ca can be readily mobilized from the skeleton to restore plasma levels unless the patient is very poorly perfused. During the period of storage the K content rises and may reach 25 mmol l^{-1} at the end of 3 weeks. Following infusion of stored blood the citrate is metabolized over the next 24 hours to bicarbonate. Hence despite the low pH of stored blood a metabolic alkalosis is a common finding on the following day. The depleted 2,3 DPG levels are restored over the next 24–48 hours. This is very important as any pre-operative blood transfusion that may be required should be completed 48 hours before surgery is contemplated.

Blood which is more than 4 days old is unlikely to contain any useful platelets. Indeed after 24 hours the platelet count of stored blood is very much reduced (50–60% of the original level). Platelet concentrates are the best way of dealing with a severe reduction in platelet count and fresh frozen plasma is the best source of clotting factors.

Although stored blood contains inadequate platelets, factor V or factor VIII, it does contain very many unwanted particles. These microemboli are made up of platelets and white blood cells which readily pass the standard blood giving set filters and lodge in pulmonary capillaries. Initially (2–7 days) the microemboli consist of platelets. Subsequently degenerate granulocytes and more platelets are added and later on fragmented red blood cells. The number of microaggregates increases with the duration of storage, to 140 000 ml^{-1} or 70 million per unit of blood. Their size ranges from 100–200 μm and they therefore pass through the standard administration set clot screen.

Platelets clump more readily in CPD blood which therefore contains more microaggregates during the first week than ACD blood. After this first week there is no difference between ACD and CPD blood in this respect. The increased acidity of ACD blood seems to prevent platelet aggregation.

If 20% or more of a patient's blood volume is replaced with transfused blood than an increase in pulmonary arteriovenous shunting and alveolar to arterial oxygen difference occurs. This may be prevented by filtration through a micropore filter (40 μm) of all blood which is more than 4 days old. Even with the use of ACD or CPD stored whole blood the indications for micro-filtration are only to prevent arterial embolus in bypass circuits and in some methods of preparation of granulocyte free transfusions. Many febrile transfusion reactions are due to granulocytes; some 11×10^8 leucocytes are necessary for such a reaction to occur. Granulocyte filtration reduces the number of these febrile reactions. It is also likely that many of the adverse pulmonary reactions to blood transfusion are caused by proteins and white cell fragments which are still in solution in plasma and are therefore able to pass through microfilters. Currently accepted indications for filtration are; cardiopulmonary bypass, transfusion of more than 5 units of blood or more than half the recipient's calculated blood volume. Packed cells are difficult to filter and contain mainly large particles which are removed by conventional filters. Platelets must never be filtered or they will be largely lost to the patient.

In addition there is a risk of hepatitis when stored blood is given and pyrogenic, haemolytic and allergic reactions may occur (Table 11.3).

Supplies of donated blood are limited in Britain and there is an increasing tendency to process blood to separate it into fractions to be used for specific purposes. The advent of specific component therapy will render microfilters redundant. If a low haematocrit is present packed red cells should be used. The plasma can then be used for the preparation of cryoprecipitate, for factor VIII replacement, fresh frozen plasma or platelet concentrates. If a massive blood transfusion is required then fresh frozen plasma containing clotting factors and platelets will be needed in addition. Rarely Ca may need to be given in situations of very low blood pressure and cardiac output when the skeleton will be inadequately perfused to maintain plasma Ca. The high citrate in stored blood chelates Ca, so massive transfusion (>30 ml kg^{-1} hr^{-1}) will require Ca supplementation. During cardiopulmonary bypass with a citrate prime there is a marked reduction in ionized Ca.

The main purpose of transfusing blood is to restore to normal the oxygen carrying capacity of the intravascular volume. There is evidence that blood flow and oxygen delivery are optimal at a haemoglobin concentration of 10 g dl^{-1}.

Currently there is rarely, if ever, any need to give whole blood. Concern about the acquired immunodeficiency syndrome (AIDS) has resulted in a change in transfusion practice in the USA and will cause us to justify our transfusion practice in this country. It is important to remember that cross matching does nothing to reduce the incidence of viral transmission, anaphylaxis or delayed haemolytic reactions.

The accepted lower limit of haemoglobin at 10 g dl^{-1} for surgery represents a 25% reduction in oxygen carrying capacity and should not be abandoned. Alternative techniques which use blood from relatives, autotransfusion and deliberate haemodilution to a PCV of 35% seem likely to be developed further. Haemodilution is particularly relevant in polycythaemia and autologous transfusion in paediatrics. If haemodilution is used the

Table 11.3   Hazards of blood transfusion.

---

1 Infection*
   viral hepatitis
   syphilis
   malaria
   cytomegalovirus
   bacterial infection after collection.
   acquired immunodeficiency syndrome
Blood should never go back into the refrigerator after 30 minutes at room temperature and packed cells should be used within 6 hours of preparation.
2 Incompatible blood transfusion
   red blood cells
   leucocytes
   platelets
   proteins
3 Air embolus
4 Allergic reactions
5 Non-immunological reactions
   vasoactive substances
   cold blood
   citrate toxicity
   potassium toxicity
   hyperammonaemia: 3 week old stored blood may have an ammonia content of
     900 ug dl^{-1}.
6 Microaggregates
7 Thrombophlebitis
8 Haemosiderosis

---

* Voluntary donors in this country are screened for syphilis, etc. and blood is taken under strict control. This is not the case in the USA where commercial donors have a higher incidence of undesirable illness.

patient should be kept normovolaemic. Viscosity falls with PCV and blood flow rate increases to compensate. The technique depends therefore on maintaining normal blood volume with plasma expander. There is no significiant disturbance of clotting or altered incidence of deep vein thrombosis.

Autotransfusion can be accomplished in a number of ways: blood can be electively removed preoperatively, stored and retransfused at a later date; it can be removed immediately pre-operatively, the patient haemodiluted and the blood transfused later in the operation; the shed blood can be salvaged pre-, or postoperatively, processed and retransfused.

The search continues for oxygen carrying fluids for blood replacement. Haemoglobin solutions and gas solvents have been most investigated rather than oxygen binding chelates.

Haemoglobin solutions These can be stored in a dehydrated form. At present their use is limited by insufficient intravascular reten-tion and increased oxygen affinity of the free Hb rather than intraerythrocytic Hb. There is potential for improvement if a tetrameric molecule is maintained. An ideal blood substitute should exert some colloid osmotic pressure to maintain intra-vascular volume, should not increase blood viscosity or cause aggregation. Stroma free Hb ($70 \, g \, l^{-1}$) has similar viscosity to plasma, potential oxygen carrying capacity and no propensity to cause aggregation or haemolysis. Rapid renal clearance occurs. Cardiac output does not increase in acute anaemia associated with isovolaemic exchange with Hb solution. However, a lower oxygen affinity solution is required without the transient renal disturbance encountered currently. Usually a Hb diuresis occurs with loss of intravascular volume. Free Hb does appear in the renal tubular cells and the lumen of the distal tubule. However, it clears within 24 hours.

Perfluorocompounds. Since oxygen is poorly soluble in water, nonpolar hydrophobic liquids such as perfluorocompounds were tried and lead to the survival of 'bloodless' rats until they re-generated their own blood cells and proteins again. This was achieved without evidence of oxygen toxicity when breathing oxygen enriched air. These compounds are chemically inert, not metabolized and some stay in the tissues a very long time whilst others leave rapidly through the lungs. The currently available Fluosol DA 20% uses an emulsifier and hydroxyethyl starch as an oncotic agent. Most of the work on this substance has been done in Japan. It has been used in the USA as a plasma expander and oxygen carrying fluid in severely anaemic (Hb 1.9–7.5 g dl^{-1}) bleeding patients who refused to accept blood. A dose of 20 ml kg^{-1} was given over 3–4 hours. With the patients breathing a low concentration of oxygen the solution carried a small amount of oxygen but breathing 100% oxygen it carried 0.8% by volume and this increase accounted for 7% of the patient's arterial oxygen delivery and 24% of oxygen consumption. At ambient oxygen tension Fluosol acts as a plasma expander but at higher inspired oxygen tensions it contributes to oxygen delivery. The oxygen is readily available because of the higher gradient for diffusion and the small size of the particles which therefore easily traverse small capillaries. Fluosol may be useful to minimize the effects of ischaemia on the heart and brain. No binding sites are involved so oxygen can be carried in the presence of other gases such as carbon monoxide. Sickle cell crisis should respond and Fluosol may be of value in extracorporeal circulation, drug overdose or fulminating hepatitis requiring total body washout techniques. Cross matching is not required.

PLASMA

The preparation of plasma now available is human albumin solution (HAS) 4.5%. This is an isotonic solution containing in each litre:

Protein 43–47 g l^{-1}.
130–160 mmol Na.
<2 mmol K.
Citrate 4–10 mmol l^{-1}.

It is a sterile solution and because it is heated to 60 °C for 10 hours during preparation it does not carry a risk of hepatitis. This solution is invaluable for replacing the plasma loss from burns and for treating crush injury and pancreatitis. In these situations the haematocrit will be raised and serial estimations will be a good guide to the volume of HAS required. If such plasma losses are replaced with crystalloid a fall in plasma colloid osmotic pressure will occur with increased formation of interstitial water and pulmonary oedema. HAS is an acid solution with pH 7.0.

Several blood components are associated with adverse reactions; factor VIII concentration may precipitate anaphylaxis.

DEXTRAN

A typical solution for plasma substitution is 6% dextran 70 which is available in 5% dextrose or 0.9% saline. It has a molecular weight (MW) of about 60 000–70 000 compared to the MW of albumin at 66 000. Dextran 70 is a hypertonic solution. Transfusion of 500 ml results in almost twice this volume of plasma expansion. In normovolaemic patients the plasma expansion lasts for 4–6 hours but in hypovolaemia expansion lasts longer. 40–70% is excreted in the urine over 24 hours.

Indications for the use of dextran.
1 Hypovolaemic shock
2 Normovolaemic haemodilution in surgical patients.
3 Prevention of thromboembolism associated with surgery.
4 Improving blood flow in a limb with a compromised circulation.

Dextran 40 is the preparation favoured for reducing blood viscosity. Its action is too short to be a useful plasma expander. Dextran 110 is also available. This preparation remains in the circulation longer than the other two.

Contraindications to the use of dextrans are the presence of cardiac failure where volume overload may occur and the presence of a bleeding diathesis, in particular a deficiency of platelets. Large quantitites of administered dextran interfere with the crossmatching of blood so it is important to take blood for this purpose prior to administering a dextran infusion. Dextran is antigenic and adverse reactions do occur including severe anaphylaxis. In one large series the incidence of shock, cardiac and/or respiratory arrest was found to be:

0.003% for plasma protein solutions.

0.008% for dextran.

0.038% for gelatin solutions.

The incidence of adverse reactions to blood transfusions is 1–6%.

With regard to the plasma expanders, the main action is anti-genicity of the large molecules in their own right and formation of soluble immune complexes. Dextrans react specifically with anti-bodies possibly produced by previous exposure to bacterial polysaccharide. Complement activation occurs in severe reactions. Haemaccel produces direct histamine release from mast cells.

GELATIN

Haemaccel is a 3.5% solution of degraded gelatin with a MW of 30 000–35 000 and a pH of 7.25. Because of the lower MW it remains in the circulation for a shorter time and its haemodynamic action is limited to about 3 hours. These solutions are quickly excreted by the kidneys, and up to 2500 ml may be given in 24 hours if necessary, whereas excessive doses of dextran, especially dextran 40, damage the kidney.

HETASTARCH

An early solution of hydroxyethyl starch (HES) was withdrawn because of residual toxic ethylene oxide remaining after manufac-ture. The newer high molecular weight solutions are partly taken up by the reticuloendothelial system (10%) but at least 25% is retained by the body for up to 72 hours. There is some question that reticuloendothelial blockade may reduce resistance to shock. Three solutions are commercially available; high MW HES 450/0.7, medium MW HES 200/0.5 and low MW HES 40/0.5. Smaller molecules are excreted in the urine and faeces. The degree of hydroxylation determines degradation by natural amylases. The incidence of anaphylaxis is about 0.006%. HES does not cause histamine release but does cause anaphylactoid reactions similar to the gelatins. Its half life is longer than that of the dextrans. There is no effect on renal function but bleeding time may be increased with impaired haemostasis and decreased factor VIII. A 6% solution of average MW 450 000 is not fully excreted for 2 weeks. There is a clinically insignificant increase in amylase and bilirubin. The total maximum dose is 1500 ml daily (20 ml kg^{-1}) or less if there is renal impairment.

CRYSTALLOID SOLUTIONS

Electrolyte solutions may be indicated for ECF losses but will *not* replace intravascular volume as they are rapidly lost into the interstitial fluid unless given in a volume 3–4 times the blood loss. These solutions diffuse throughout the ECF and with a large crystalloid load oedema accumulates in muscle, subcutaneous fat, skin, myocardium and intestine.

If the PCWP is raised in the presence of shock then vasoactive or inotropic drugs, for example sodium nitroprusside, nitroglycerine or dobutamine may be indicated but further discussion of these is beyond the scope of this book.

Pulmonary oedema will occur if the PCWP is greater than 15–20 mmHg or at a lower value than this if the colloid osmotic pressure is reduced or if the pulmonary capillaries are excessively permeable as in adult respiratory distress syndrome (ARDS).

In these circumstances the gradient (colloid osmotic pressure — PCWP) is valuable since it correlates with the absorptive force tending to keep the alveoli free of oedema fluid.

Endorphines have been implicated recently in the pathophysiology of shock. Naloxone can increase the blood pressure and survival in experimentally induced hypovolaemic shock.

Other preoperative fluid losses should be replaced slowly over 24–48 hours and adequate i.v. therapy given to cover continuing losses.

Choice of fluid for resuscitation from severe shock remains controversial. Colloids are in general preferable as judged by oxygen debt, volume required and evidence of pulmonary overload. A mixture of crystalloid and colloid is probably optimal. Regimens with at least some colloid included produce greater increases in haemodynamic and oxygen transport variables after albumin than after Hartmann's solution given in 2–4 times the volume. Where pulmonary capillary permeability is damaged as in the adult respiratory distress syndrome, oncotic pressure should be maintained at a minimal critical level. In the normal person, however, efficient protective mechanisms prevent interstitial pulmonary oedema even with large crystalloid infusions.

In experimental animals interstitial volume was restored more rapidly in resuscitated awake animals than in anaesthetized animals. Early metabolic changes after haemorrhage may result in an important increase in solute production.

NORMAL MAINTENANCE REQUIREMENTS

Water   0–10 kg   100 ml kg^{-1} daily
        11–20 kg   1000 ml + 50 ml kg^{-1} daily
        21 kg or more   1500 ml + 25 ml kg^{-1} daily
  For an adult this is about 30–35 ml kg^{-1} daily.
Sodium   1–2 mmol kg^{-1} daily.
Potassium   1 mmol kg^{-1} daily.

Certain other aspects of preoperative fluid therapy include administration of dextrose and insulin to diabetics and setting up a mannitol infusion in patients with obstructive jaundice to reduce the incidence of acute tubular necrosis.

INTRAOPERATIVE FLUID AND ELECTROLYTE REQUIREMENTS

There are a number of factors to be considered.

The metabolic response to trauma

This results in the following changes:
1   Urine volume is reduced and its Na content is reduced.
2   Urine K and nitrogen (N_2) content are raised.
3   Plasma Na and albumin are decreased.
4   Plasma ADH, aldosterone, cortisol, catecholamines, insulin, glucose, free fatty acids (FFA) and amino acids (AA) are increased.

Intraoperative factors

Vascular permeability is increased in damaged areas giving rise to increased interstitial fluid. This sequestration into a third space was first described by Shires and his colleagues in the 1960s. It results in a reduced plasma volume. The existence of this third space is disputed; most authorities agreeing with Cleland (1966) that the ECF reduction was overestimated. In some circumstances for example crush injury and bowel obstruction large volumes of fluid are sequestrated but otherwise the importance of the third space is still subject to uncertainty. ECF expansion after intestinal anastomosis increases oedema at the anastomotic suture line which if it persists could be deleterious.

Fluid administration

During major surgery a solution such as Hartmann's balanced electrolyte (see below) which resembles ECF is often administered. Urine volume and Na excretion are thereby maintained with less intraoperative hypotension and less postoperative renal failure. However, over enthusiastic administration of Hartmann's solution may expand the interstitial fluid especially in the pulmonary circulation leading to pulmonary oedema.

Table 11.4 Composition of intravenous fluids (mmol 1^{-1}).

	Na	K	Cl	Ca	Lactate
Hartmann's	131	5	111	2	29
Dextrose + saline (4.0% + 0.18%)	31		31		
0.9% saline	150		150		

Renal impairment

Postoperative renal failure is commoner after cardiopulmonary bypass, aortic or renal vessel surgery and with pre-existing hepatic or renal disease or when hypotension, massive trauma and transfusion have occurred. General anaesthetics also may impair renal function.

Fluid replacement

Many operations do not require fluid administration. The volume and composition of fluid required depends upon a number of factors.

1   The pre-existing abnormality which requires correction.
2   The maintenance requirements.
3   The extent of the operation site oedema.
4   The presence of exudates and intestinal secretions.
5   The blood loss.
6   The extent and duration of the operation.

The aim is to give fluid to replace sequestered loss which is very difficult to measure, and to supply basal requirements. Successful treatment also reduces viscous bronchial secretions post-operatively and reduces the sensation of thirst.

Therefore, Hartmann's solution may be given to maintain a urine volume of $0.5–1$ ml kg^{-1} hour^{-1}. An alternative is to give 10 ml kg^{-1} for the first hour and 5 ml kg^{-1} during subsequent hours to a maximum volume of 3 litres. Other workers recommend 5 ml kg^{-1} hour^{-1} of surgery to a maximum of $1.5–2$ litres in an adult with a reduction for pre-existing cardiovascular disease. Hartmann's solution is slightly hypotonic in relation to interstitial fluid and therefore will replace the fluid lost during surgery and the insensible water loss due to sweating and respiration. Administration of dextrose saline and 5% dextrose may cause hypotonicity. There is a danger in excessive administration of salt solutions and especial care should be taken *not* to increase the rate of infusion to compensate for a low blood pressure (BP) due to another cause, for example anaesthetic drugs. If, despite adequate intraoperative fluid administration and in the absence of outstanding loss, oliguria persists then 20–40 mg frusemide should be given i.v. Blood loss in excess of 1 litre in an adult requires replacement but losses of less than 1 litre can be replaced with electrolytes. The haemoglobin concentration falls but oxygen carrying capacity is not significantly reduced. Patients with persistent gastrointestinal bleeding and profound anaemia may continue to bleed until near normal haematocrit is restored by packed cells. Haematocrit affects viscosity and therefore blood flow with a marked increase in bleeding at low viscosity.

There are many circumstances where blood loss is difficult to estimate especially if large volumes of irrigating fluid have been used such as in transurethral and neurosurgery. A cell count can be done on the irrigating fluid provided it is almost isotonic.

Colloids

These should be reserved for expansion of a reduced intravascular volume when blood is not available. Excessive use of colloid leads to acute circulatory overload because it does not enter interstitial fluid. CVP measurements are valuable during colloid administration.

Special circumstances

Cardiovascular disease

There is a danger of overexpansion of the ECF space. A urinary catheter and CVP may be indicated preoperatively and perhaps inotropic drugs or diuretics. In addition, the importance of K replacement in those receiving digoxin and diuretics cannot be overemphasized. A patient should not be presented for surgery with a significant reduction in total body K.

Renal insufficiency

In chronic renal failure fluid depletion may further increase blood urea and creatinine. Dialysis should be manipulated to produce an optimal state for surgery. Methods of protecting the kidney during surgery will be considered in Chapter 13.

Hepatic disease

The dangers include development of renal failure, fluid overload especially if ascites is already present and hypoglycaemia.

If the patient has obstructive jaundice, mannitol or frusemide should be administrered to prevent renal failure. (See Chapter 14.)

Prophylaxis of deep vein thrombosis

Dextran given by the following regimen has been shown to be as effective as low dose heparin in preventing fatal pulmonary embolus.

Dextran 70 in 0.9% saline, 500 ml:
1 At induction of anaesthesia over 1 hour.

2 Postoperatively over 2 hours.

3 On the first postoperative day over 4 hours.

Recent evidence suggests that blood clots faster when diluted with saline (Janvrin *et al.* 1980). Sixty patients undergoing laparotomy were allocated to two groups, only one of which received intra-operative fluid therapy. Those receiving parenteral fluid became hypercoagulable compared to the group not receiving fluid. Post-operative deep vein thrombosis occurred in 30% of the group receiving i.v. fluids (as measured by ^{125}I fibrinogen uptake test) but in only 7% of the group from which fluids were withheld.

POSTOPERATIVE FLUID THERAPY

At this time the patient has the following problems to contend with:

1 The metabolic response to trauma.

2 The third space effects.

3 The effects of anaesthesia and surgery on renal function.

4 The well meaning but puzzled doctor.

Important factors are maintenance of hydration, renal function, blood volume and electrolyte balance. The metabolic response to trauma has been fully considered elsewhere and leads to retention of Na and water. Third space ECF losses continue from the operative into the postoperative period.

Postoperative renal function

After surgery a reduction in urine output is to be expected (Table 11.5). In a stable patient with a given osmolar load for excretion at a fixed GFR and urine flow an increase in distal tubular water reabsorption will reduce urine volume. Another mechanism operating after surgery is the presentation of an increased osmolar load for excretion. This may be due to protein catabolism and in the absence of an increased free water excretion will increase blood urea.

It is important to know what may be considered abnormal. A urine volume of less than 20 ml hourly for two or more hours should be considered abnormal because this is less than the minimum urine volume formed with maximum water conservation and minimum solute load. It is, however, not uncommon for patients

Table 11.5 24 hour urine measurements in normal and postoperative states.

	Normal	Post-op	Duration days
Volume	1500 ml	500 ml	1–2
Sodium	70 mmol	5–20 mmol	2–5
Potassium	70 mmol	100 mmol	1–3
Nitrogen	8–12 g	10–20 g	4–10

to be almost anuric intraoperatively and postoperatively and subsequently to return to the normal urine flow. Before embarking upon therapy for oliguria it is essential to check for urinary obstruction.

Unfortunately, a urine flow in excess of 20 ml hour^{-1} does not exclude renal dysfunction. Severe intercurrent disease, hypotension and renal disease may present with a urine flow greater than 20 ml hourly in the presence of a rising blood urea. This is probably due to a decreased functioning nephron mass so that total urea excretion is small and urine urea low.

The importance of maintaining renal perfusion intraoperatively and postoperatively cannot be overemphasized. Although the controversial third space losses may in the past have been overestimated with detrimental pulmonary oedema, underestimation is just as dangerous since reduction in ECF and plasma volume will intensify intra- and extrarenal mechanisms for Na and water conservation. Although overexpansion of ECF increases Na excretion only a proportion of the extra administered Na load is excreted. (The fraction of the load eliminated remains constant at about 33% under conditions of anaesthesia and surgery.)

Summary of the changes occuring postoperatively

1 Decreased GFR.

2 Increased tubular reabsorption of Na partly due to an increase in aldosterone.

3 Increased collecting duct reabsorption of Na.

4 Increased ADH and water retention.

5 Inability to excrete free water or form hypotonic urine. There is therefore some problem with regard to which fluid to give.

There are two alternatives:

(a) A fluid load may be given to produce a normal urine output, to inhibit ADH response, maintain plasma volume and reduce renal vasoconstriction.

(b) Fluid may be restricted on the grounds that a high fluid regime has little effect on ADH production.

High salt loads are tolerated but not beneficial and a high urine output is only achieved at the expense of positive Na and water balance with ECF expansion, although hyponatraemia will be prevented. In a healthy person 2 l of 0.9% saline leads to a reduction in pulmonary compliance and continued administration in a postoperative patient may produce pulmonary oedema. In salt load resuscitation of battle casualties pulmonary oedema may have a mortality of up to 80%. Infusion of dextrose only results in hyponatraemia. After 24–48 hours a negative water balance will restore Na to normal. Severe sepsis needs more expansion of the interstitial space than is the case in haemorrhagic shock.

Postoperative requirements

Water requirements

Preoperative requirements and the adequacy of correction of deficits must always be taken into account. Some postoperative water loss may be considered obligatory including insensible loss, urinary loss and dynamic loss.

INSENSIBLE LOSSES

The volume will vary with temperature. There is an increase in insensible loss of 10% per degree centigrade rise in body temperature. This loss also varies with the humidity and factors such as breathing dry anaesthetic gases for a prolonged period. Insensible water gain may also occur due to the production of water of metabolism. This may reach high levels in massive injury, sepsis and fever and is made worse by mobilization of water from cells into ECF in an amount up to 850 ml daily. Burns lose water by evaporation. In states of water and electrolyte lack decreased insensible loss occurs and there is also wide variation between individuals from day to day.

URINARY LOSSES

The postoperative patient cannot excrete free water or produce a very hypotonic urine for 72 hours after surgery during which time urine osmolality is about 1 mosmol ml^{-1} and 300–400 ml of urine are required to clear the solute load in a starved patient. In catabolism a much larger volume of urine is needed to prevent solute retention. If no intravenous fluid is given to a patient postoperatively then he voluntarily takes about 25 ml kg^{-1} of water daily. Therefore administration of free water should perhaps be limited to this volume in the absence of increased water loss. Administration of 5% dextrose constitutes a water load which given in excess may lead to water intoxication over 48–72 hours.

DYNAMIC LOSSES

Normal gastrointestinal secretions are reabsorbed but fluid loss due to vomiting and from fistulae should be taken into account. Ileal, biliary, and pancreatic fluid losses have an electrolyte content similar to 0.9% saline with 5 mmol K in each litre (see Table 11.1). Colonic losses and those above the pylorus may alternatively be replaced with 0.45% saline with 5 mmol K in each litre. Protein loss may also require replacement. Fistulae losses, however, are usually of mixed origin and therefore should have their electrolyte content measured.

Electrolyte requirements

SODIUM

An excessive Na load requires 2–3 days for excretion. Normal Na requirements are 80–100 mmol (1–2 mmol kg^{-1}) daily but postoperatively there is virtually no Na requirement in the absence of excessive loss. Children normally require about 2 mmol kg^{-1} Na daily. Certainly in patients with pulmonary, renal or cardiac disease salt restriction is indicated unless chronic renal failure with fixed Na and water loss is present when rapid hypovolaemia may occur with such restrictions.

Potassium. Losses are 70–90 mmol daily (1–1.5 mmol kg^{-1}). Since immediate postoperative renal function is uncertain K should not be given on the first postoperative day. Thereafter 50 mmol should be given daily and this should be increased depending on losses and plasma levels. If K is not given postoperatively the tendency for Na to be retained is prolonged and intensified. Aldosterone increases postoperative urinary K loss.

Magnesium. 5 mmol are lost in the urine daily. Replacement is only important with long term parenteral and diuretic therapy.

Calcium. The vast stores in the skeleton replace plasma Ca in the immediate postoperative period.

Blood, albumin and other colloids should be used as required and are discussed elsewhere.

If fluid replacement is judged to be adequate and urine output is poor it is important where diuretic therapy is contemplated to consider the state of the cardiovascular system. Administration of frusemide results in a reduction in cardiac output and a fall in systemic and pulmonary arterial pressures. Mannitol on the other hand causes an increase in cardiac output and pulmonary arterial pressure. Therefore, in treatment of postoperative oliguria in the absence of known cardiovascular disease mannitol is the diuretic of choice.

Planning of postoperative fluid regime

The nature of the surgery and the current fluid balance of the patient are of paramount importance here. It is known that 3 litres of water daily puts most patients into marked positive water balance but 1.5 litres of 5% dextrose with no added salt very rarely produces hyponatraemia in the absence of large abnormal salt losses. One litre of 0.9% saline daily leads to weight gain and Na retention. In the postoperative period hyponatraemia is almost always due to administration of excessive water as 5% dextrose.

A slightly dry but stable patient is the aim. Currently overhydration is more of a problem in surgical patients than underhydration especially in the elderly, after trauma and intracranial surgery. Elective surgery of moderate severity can be accomplished without i.v. fluids and the patients come to no harm.

The following recommendations can be made:

1 Give blood to keep the haemoglobin greater than 10 g dl^{-1}.

2 Maintain a urine output in excess of 400 ml daily. A greater urine output should be aimed for if there is an increased solute load for excretion.

3 1.5 litres of 5% dextrose over 24 hours will replace insensible loss, reduce N_2 loss and maintain urine output.

4 No Na need be given for 3 days if there is no loss. This situation is rare so that a compromise is to give dextrose 4.0% saline 0.18% if minimal ECF losses are occurring. This contains Na 31 mmol l^{-1}. Hartmann's solution is more appropriate if ECF losses are raised.

5 After the first day 1 mmol kg^{-1} of K should be given daily as KCl.

6 Plasma albumin should be maintained in excess of 35 g l^{-1}.

7 Abnormal gastrointestinal losses should be replaced.

Mg, vitamins and zinc are required during prolonged i.v. therapy and parenteral nutrition should be considered as suggested in Chapter 12. Intravenous fluids should be continued until oral fluids can be taken in adequate volume. Great care should be taken not to overload the patient at this time. As oral intake is increased a reduction in i.v. fluids is appropriate. Increasingly, use is being made of a programmable calculator for fluid requirements depending on the patient's weight, age and predetermined requirements. This is held to reduce the incidence of complications and avoid overtransfusion whilst supplying estimated requirements.

Examples

1 Cholecystectomy.
Intraoperatively. 1 litre Hartmann's solution for ECF replacement.
Postoperatively. 1.5 litres 5% dextrose and 0.5 litres of Hartmann's solution for ECF replacement over 24 hours.

2 Major surgery involving prolonged exposure of a body cavity, for example, thoracoabdominal oesophagectomy or aortic aneurysm repair.

Intraoperative

(a)   1–3 litres Hartmann's solution for ECF replacement. The volume will depend on CVP and haematocrit measurements.

(b)   250 ml hour^{-1} 5% dextrose will cover evaporative losses due to visceral exposure.

Postoperatively. 1.5–2 l 5% dextrose and 0.5–1 litre Hartmann's solution for ECF replacement over 24 hours.

Chapter 12
Nutrition

Critically ill patients may show a marked stress response to trauma and disturbances of carbohydrate, fat and protein metabolism. Such patients show increased oxygen consumption, fluid overload and susceptibility to sepsis. In a group of patients with severe complications after abdominal surgery, those who lost more than 30% of their original weight died. It has been estimated that 40–50% of medical and surgical patients will show protein energy malnutrition at some time during their hospital stay. Unfortunately we do not have a simple nutritional index of sufficient predictive power to define a group of patients who need feeding.

Some nutritional disturbances have far reaching effects such as disturbances in drug disposition, neurotransmitter release and conscious level. Several of these effects can be modified by various forms of nutritional supplementation, although water retention resulting from increased aldosterone and antidiuretic hormone secretion (ADH) often limits the volume of nutritional support possible.

Complications of overzealous nutritional replacement can be severe, ranging from respiratory failure to hyperosmolar states. An understanding of basic metabolism and energy requirements is therefore very important. It is illogical to infuse nutritional substances without knowledge of energy requirements and metabolic measurement of the consequences.

Table 12.1 Nutritional reserves of a 70 kg man.

	kg	Duration	kcal
Carbohydrate (mainly liver glycogen)	0.2	6–12 h	800
Fat	12–15	20–25 days	109000–136000
Protein (mainly muscle)	4–6	10–12 days	16000–240000

Stress response to trauma

Surgery or trauma activates a typical metabolic and hormonal response which may be mediated by pain or some other neural mechanism and may be accompanied by immunosuppression. Cuthbertson and Tilstone described an 'ebb' and 'flow' phase. Loss of nitrogen (N_2), potassium (K) and sodium (Na) from the body are greatest during the flow phase and changes recede if the patient recovers and anabolism begins.

Classicially, the metabolic response to trauma is divided into four phases:

1 The phase of injury. This lasts 2–4 days after major trauma unless it is prolonged by sepsis, continued fluid loss or acid base disturbance, to about seven to ten days. Changes which occur are due to adrenergic and adrenocorticoid hormones and represent catabolism. The purpose of this phase is to allow flight to safety and to maintain blood volume.

2 The turning point. The effect of phase 1 hormonal activity is terminated.

3 Anabolism. This phase begins with positive nitrogen balance and an increase in muscle mass. It may last up to 5 weeks after major trauma.

4 Fat gain. This occurs over several months and involves replenishment of the fat stores that were present before injury.

Consequences of major operation or injury

1 There is direct destruction of cells in the wound.

2 There is loss of fluid into and from the damaged area.

3 Operation and injury interfere with food intake, so that the effects of starvation are added.

4 An increase in energy expenditure occurs as metabolism increases to deal with tissue destruction and repair damage.

5 Hormonal changes occur which have both defensive and restorative actions.

In rats division of afferent fibres from the site of injury suppresses development of the neuroendocrine response. Extensive epidural analgesia to block both somatic and sympathetic afferent activity, centrally acting morphine, fentanyl, alfentanyl and high spinal cord lesions also suppress the response which can be modified by pre-existing fear and anxiety. There is some evidence for a

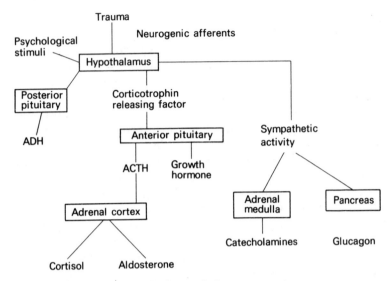

Fig. 12.1 Hormonal changes in the metabolic response to trauma.

humoral pathway, which would not be modified by the above measures; prostaglandins, bacterial endotoxin and other pyrogens have been postulated as 'wound hormones'. Interleukin-1 is the best investigated wound hormone which increases lysosomal protein degradation, an action which is mediated by prostaglandin E2 although other peptides are also concerned.

Endocrine changes

The hypothalamus acts as a final common pathway for the neuro-endocrine response which involves secretion of pituitary hormones and sympathetic activity (Fig. 12.1). Most hormones increase in proportion to the severity of the surgery although ACTH is a poor index of the severity of injury and the growth hormone response may vary with age.

Insulin secretion is depressed during surgery, presumably due to increased noradrenaline. Insulin levels can be increased by alpha block but suppression is shortlived in uncomplicated surgery returning to normal in about 7 days. Plasma glucose concentration and beta-adrenergic activity increase. There is also evidence of tissue resistance to insulin. Early hyperglycaemia (6–18 hours) is

Table 12.2 Hormones involved in endocrine response to surgery.

Neuroendocrine response	Systemic response
ACTH	Insulin
Vasopressin	Cortisol
Growth hormone	Glucagon
TSH	Thyroxine, triiodothyronine
Adenaline	Aldosterone
Noradrenaline	Angiotensin

probably the result of adrenal medullary output rather than sympathetic nervous discharge. In patients undergoing cholecystectomy there is no difference between 0.9% saline and Hartmann's solution in their effect on glucose and insulin but infusion of 5% glucose exacerbates the hyperglycaemic response and increases plasma insulin concentration suggesting that the usual suppression of insulin secretion during abdominal surgery can be overcome.

Adrenaline and noradrenaline levels are raised from 24 hours up to 3 days after trauma. Potent stimuli causing secretion of these hormones are sepsis, hypoxia, shock and fear. Very high plasma adrenaline levels occur in stressed conditions such as cardiac arrest when it stimulates the renin-angiotensin system (beta effect), and the pituitary causing release of ACTH. These serve to amplify the effect of the catecholamines by further increasing their release. In response to more mundane stress the sympathetic nervous system is more important with high concentrations of noradrenaline in the synaptic clefts. Whether such high levels of adrenaline as occur in flight or fright reactions are beneficial is hard to establish. High serum levels of catecholamines in shock are probably responsible for the lactic acidosis and initial hyperglycaemia.

The increase in cortisol is rapid, occurring within a few hours and is in proportion to the severity of surgery. Cortisol is termed the permissive hormone since it has little effect on glucose production but by stimulating peripheral release of amino acids (AA) which act as substrates it will increase gluconeogenesis. Opiate inhibition of the pituitary adrenal axis is mediated via a noradrenergic pathway (administration of naloxon produces a rise in plasma cortisol).

Glucagon rise is also related to the severity of injury and reaches a peak 18–48 hours after injury. Increased glucagon is thought to be due to increased catecholamines. Gluconeogenesis, glyco-genolysis and AA uptake are stimulated in the liver.

Changes in thyroid hormones are variable; there is a post-operative peak in TSH, thyroxine is little changed, tri-iodothyronine falls and reverse tri-iodothyronine rises. Epidural analgesia which suppresses cortisol has no effect on the thyroid hormones. Changes may be secondary to increased oxygen consumption after surgery. Growth hormone has only a minor effect; a normal metabolic response to trauma occurring in hypophy-sectomized patients receiving steroid replacements. These hormonal effects mostly stimulate glucose production and catabolic processes; the so-called 'catabolic drive'.

ADH rises rapidly during surgery, in response to neuronal stimulation such as traction on the mesentery and in uncomplicated cases remains elevated for 2–4 days after surgery. This response to trauma overrides the normal regulation of ADH by ECF osmolality and may result in inappropriate antidiuresis.

There is some opiate control of the pituitary-adrenal axis; infusions of met-enkephalin analogues reduce circulating cortisol despite a reduction in blood pressure. Naloxone produces a pronounced increase in adrenocorticotrophic hormone, cortisol and lipotrophin. Constant tonic inhibition of the pituitary axis throughout 24 hours is postulated. There is an increased demand for glucose after surgery to provide energy for wound repair. Whilst glycogenolysis continues, stimulated by catecholamines and glucagon, there is also an increase in lactate and pyruvate release from muscle. Pyruvate is taken up by the liver as a gluconeo-genetic substrate.

In severe injury whole body protein turnover increases but breakdown increases more than synthesis. There is a complex interaction between nutritional state and the degree of injury in severe trauma, burns and sepsis and protein breakdown is relatively resistant to nutritional modification. Catecholamine infusions used in these circumstances further increase glyco-genolysis and hepatic gluconeogenesis and worsen insulin resistance. Fat breakdown increases the release of glycerol and FA which may be taken up by muscles and used directly as an energy source or by the liver for conversion into ketone bodies. Adrenaline and glucocorticoids stimulate adipose tissue lipolysis whilst glucagon speeds intrahepatic conversion of fatty acids (FA) to ketone bodies.

One of the most prominent features of the stress response is therefore hyperglycaemia; both glucose oxidation and turnover are increased and the primary disturbance is one of increased hepatic output of glucose. Hyperglycaemia is a common finding in patients on admission to hospital with conditions such as myocardial infarction. Administration of 60 g dextrose to labouring women significantly increases maternal dextrose and umbilical dextrose and insulin. Glucagon and arterial pH are reduced. The increase in insulin results in neonatal hypoglycaemia.

The effects of resistance to insulin after injury are as follows;
(a) the maximal rate of glucose disposal is reduced.
(b) metabolic clearance rate of insulin is almost twice normal.
(c) peripheral tissue such as muscle are insulin resistant.

Severe metabolic derangements occur in the septic state with increased glucose utilization and often inappropriately low insulin which correlates with cardiovascular decompensation, reduced cortisol and reduced growth hormone both of which are associated with a high mortality. Recovery is associated with return of a normal reciprocal relationship between insulin and glucagon. Total body oxygen consumption is increased. This may be due to the hypermetabolism of infection, catecholamine induced after injury or following administration of excess glucose during parenteral nutrition.

FACTORS WHICH MODIFY THE EXTENT AND DURATION OF METABOLIC RESPONSE TO TRAUMA

1 Pain and analgesia.
2 Fear and sedation.
3 Post-trauma complications;
(a) Shock.
(b) Haemorrhage.
(c) Hypoxia.
(d) Sepsis.
4 Pre-existing nutritional status and debility.
5 The extent and site of the trauma.

In addition to these factors, there is a considerable individual variation in response and postoperative cortisol rise is greater in women than men. After major, uncomplicated surgery, such as gastrectomy, the energy expenditure may rise to 3500 cal daily for 4 or 5 days. The addition of sepsis may increase this expenditure to 5000 cal daily. After an uncomplicated gastrectomy weight loss may amount to about 3 kg. Half of this will be from fat and half from protein. It is interesting that in animals the normal response to trauma is to lie quite still in a warm environment. This minimizes the need for increased heat production and reduces oxygen consumption and carbon dioxide production. In a more extreme situation nitrogen losses may amount to 35 g daily which corresponds to in excess of 200 g protein or 1200 g of lean tissue mass (muscle mass) daily. It has been shown that the postoperative increase in metabolic rate rises in direct relationship to the severity of the operation. An increase in energy consumption of about 10% may be expected after elective, uncomplicated surgery. This may rise to 20–50% in association with peritonitis and up to 100% after severe burns. In the absence of nutritional support, this additional energy is derived from the fat stores and increased breakdown of tissue protein.

METHODS OF MINIMIZING THE METABOLIC RESPONSE TO TRAUMA

This response can, to some extent, be obtunded by good surgical technique, with minimal tissue damage and good intraoperative management. There should be a minimum delay between the time of trauma and medical and surgical treatment, as any delay prior to definitive treatment has a significant effect on morbidity and mortality.

1 Blood and fluid losses should be precisely replaced.

2 Perfect oxygenation should be maintained under all circumstances.

3 The patient should receive adequate nutritional support.

4 Adequate analgesia should be given at all times.

There is apparent augmentation of lymphoctye responsiveness in some patients after premedication which suggests that relief of pain and anxiety may reduce hormone levels and allow lymphocytes to react effectively. It has been recognized for some time that afferent sensory impulses from the injured area are important factors in precipitating these hormonal changes. Profound analgesia and a sensory blockade have been tried in an attempt to prevent the metabolic response to surgery, by extradural and spinal anaesthesia and large doses of intravenous analgesic agents. Extensive extradural analgesia may inhibit the cortisol and glucose response to lower abdominal surgery. However, it is much less effective in abolishing the metabolic response in upper abdominal and thoracic surgery. Although thoracic block of sympathetic fibres may prevent hyperglycaemia, the hormonal changes may well result from stimulation of vagal afferent fibres and therefore be unresponsive to extradural block. However, neuroendocrine responses to surgical stress are prevented in patients with low spinal anaesthesia who have no suppression of efferent adrenergic tone. Phentolamine modifies the glycaemic response to gynaecological surgery. An alternative approach is to use large doses of intravenous morphine or the synthetic opiate, Fentanyl. Indeed, it does seem that large doses of these narcotic analgesics will abolish the cortisol, growth hormone and hyperglycaemic response to major trauma. Fentanyl and oxygen anaesthesia prevents the ADH response in patients with heart disease undergoing surgery although has very limited metabolic benefit in the postoperative period. Although nitrogen balance can be improved for the first 5 days after surgery by such manipulation, it is not yet clear whether obtunding the metabolic response to trauma acutely, has in fact any long term benefit on aspects such as negative nitrogen balance. Most anaesthetic techniques only modify the response to surgery for the duration of their intervention and pain relief per se may have little effect.

Naftidrofuryl is a drug which inhibits intracellular metabolic pathways and when infused into patients undergoing elective operations of moderate severity significantly improves postoperative nitrogen balance. Its precise mechanism of action remains in doubt.

These metabolic effects have a profound effect on the water, salt and potassium balance.

Water balance

An increase in ADH will produce profound antidiuresis by preventing the kidney excreting free water. Characteristically, urine volume falls and urine osmolality rises, although glomerular filtration rate will be normal if the blood volume is normal.

Endogenous water formation will be increased. Fat is broken down into free fatty acids and glycerol and muscle is broken down into its constituent amino acids; both of these are to provide energy substrates for gluconeogenesis. Oxidation of 1 kg fat to carbon dioxide and water results in the formation of about 1 litre of water. Similar volumes of water are produced by complete oxidation of muscle protein. It is important to include this volume in the calculation of water balance postoperatively. In these circumstances metabolic water which is released can seldom be excreted by the kidney and is retained to diffuse throughout body fluid compartments, thereby increasing total body water. This may produce hyponatraemia. Therefore the effects of fat oxidation, protein breakdown and ADH secretion result in an increase in ECF volume and haemodilution.

Salt balance

The hormonal changes which occur result in marked Na retention. Daily urine Na may fall as low as 1 mmol. Chloride retention usually accompanies this but to a slightly lesser degree. Despite this, plasma Na falls as relatively more water is retained and dilutional hyponatraemia usually occurs. If there is any degree of hypovolaemia or reduction in ECF volume, then aldosterone secretion persists and may prolong Na retention.

Potassium balance

Potassium balance parallels nitrogen balance. 70–90 mmol of K are excreted in the urine daily if no K is given. Cell breakdown is one cause of this K loss but the K to nitrogen ratio in the urine is much greater than the ratio within cells. Some of this K which moves out of the cells will then be excreted in the urine. The increased secretion of aldosterone will result in an increased exchange of Na for K in the renal tubules.

Normally after the seventh day these hormonal changes are reversed. Na, K and water balance return to normal. The phase of anabolism will then begin.

MALNUTRITION

The effects of malnutrition

Acute weight loss in excess of 20% is associated with a post-operative mortality of 33%, compared to 3.5% in those who had lost less weight. One-third of patients with physical disease causing loss of body weight die within two years, although there may be no increase in intraoperative risk in patients who have lost weight. Malnutrition is generally held, however, to lead to progressive weakness and other well known effects listed in Table 12.3, although the efficacy of perioperative parenteral nutrition in reducing mortality and morbidity has been questioned. The ability to respond to infection is due in part to a group of proteins, leukocyte endogenous mediators, which are reduced in malnutrition and restored by increasing protein intake. The reduction in respiratory muscle strength and maximum minute ventilation may impair the capacity of the respiratory muscles to handle an increased ventilatory load in the event of thoracopulmonary disease. Nutritional support improves the antibody response to influenza virus vaccine in the elderly.

Table 12.3 The effects of malnutrition.

Progressive weakness
Reduced vital capacity, respiratory rate, minute volume
Increased risk of respiratory infection
Difficulty weaning from ventilatory support
Reduced cardiac output, myocardial contractility and compliance
Reduced tensile strength of skin, increased wound dehiscence
Breakdown anastomoses
Reduced plasma proteins, susceptibility to salt and water overload
Reduced host resistance

A reduction in protein and calorie intake reduces protein turnover, synthesis and breakdown. During repletion there is both increased synthesis and increased breakdown but the former predominates with net accumulation of protein. Starvation occurring in the previously well-nourished, increases plasma levels of branched chain amino acids (BCAA), glycine and reduces plasma alanine. Specific plasma amino acid patterns are found in sepsis and hepatic failure but these do not necessarily reflect the amino acid concentration of the muscle or the total pool of free AAs.

Nutritional assessment

Most patients requiring intensive therapy show some evidence of malnutrition although the aim in general is identification of patients with marginal malnutrition who might benefit from nutritional intervention to reduce postoperative mortality and morbidity. A great variety of clinical and laboratory parameters have been recommended in the past for evaluation of malnutrition (Table 12.4) but few are specific enough to be of much help.

Table 12.4 Clinical and laboratory values and malnutrition.

Parameter and standard	Degree of malnutrition		
	Mild	Moderate	Severe
Weight loss (of usual)	less than 10%	10–20%	more than 20%
Weight (of ideal)	80–90%	70–79%	less than 70%
Anthropometric measurements			
Triceps skinfold (of standard)	80–90%	60–79%	less than 60%
Arm muscle circumference (of standard)	80–90%	60–79%	less than 60%
Creatinine-height index (of standard)	60–80%	40–59%	less than 40%
Biochemical measurements			
Serum albumin (3.5–5.0 g dl^{-1})	3.0–4.0	2.9–2.1	less than 2.1
Serum transferrin (175–300 mg dl^{-1})	150–175	100–150	less than 100
Thyroxin-binding pre-albumin (28–35 mg dl^{-1})	25.2–28	23–25.2	less than 23
Retinol-binding protein (3.6 mg dl^{-1})	2.7–3	2.4–2.7	less than 2.4
Immune competence			
Total lymphocytes (1500–5000 mm^{-3})	1200–1500	800–1200	less than 800
Delayed cutaneous hypersensitivity	reactive	relative energy	non-reactive

It is difficult to obtain an ideal body weight in very ill patients and loss of body components is the final stage of malnutrition.

Most other methods of assessment have their limitations and plasma concentrations of visceral proteins are affected by conditions other than malnutrition such as hydration and sepsis. Total body K is reduced in Crohn's disease and parallels the degree of illness; if it is less than 70% normal nutritional support is indicated. 3 methyl-histidine was originally thought to be a valuable indicator of muscle breakdown but there is doubt about its specificity. It is, therefore, still very difficult to identify at-risk patients although the presence of two of the following three parameters gives a good indication:
1 Unintentional weight loss greater than 10%.
2 Albumin less than 35 g l^{-1}.
3 A negative reaction to five skin antigens.
The independent effects of infection and trauma also play a major part in nutritional status.

Of equal value may be a history and clinical examination taken with the haemoglobin level and acute visceral proteins.

The effect of malnutrition on immune function

Malnutrition modifies immune function. Cell mediated immunity can be assessed by response to recall skin antigens. A marked reduction in response has been found to correlate with severe sepsis and poor outcome. Although such testing has been advocated as a useful indication of malnutrition it is non-specific and routine use of delayed hypersensitivity skin testing is not very helpful.

Reliability of assessment

Hand grip dynamometry seems to be a useful screening test for detecting malnutrition. Loss of muscle power does predict those patients likely to show serious postoperative morbidity. Development of muscle fatigue is a consistent finding in malnutrition and one which is not abnormal in non-specific situations such as sepsis, administration of steroids, anaesthesia or moderate trauma.

However nutritional depletion is assessed, and loss of muscle power is at present the best guide, *all* patients require adequate nutrition ideally administered by the enteral route. If this is not possible then nutrition will have to be administered parenterally. Early enteral feeding is thought to reduce secretion of catabolic hormones and return immune responsiveness more rapidly.

Indications for parenteral nutrition (PN)

Indications include: acute hypercatabolism (multiple trauma, burns, septicaemia); pyloric stenosis; pancreatitis; cardiac surgery; gastrointestinal cutaneous fistulae; inflammatory bowel disease; cancer surgery and cachexia. Many of the conditions requiring preoperative PN will need continued feeding into the post-operative period.

Assessment of requirements for parenteral nutrition

A variety of techniques exist for investigating substrate metabolism in patients. These include indirect calorimetry, substrate load tests, measurement of arterio-venous differences (Fick principle) and isotope infusions.

The measurement of gas exchange, although seldom clinically available, can be useful for evaluation of the nutritional needs of hospitalized patients. If the resting respiratory quotient (RQ) is known together with nitrogen excretion, then the proportion of calorie requirements to nitrogen can be calculated.

Measurement of energy requirements must be expressed as a function of body size. Measurements of total body energy expenditure in healthy subjects at rest, represents a basal metabolic rate of total cell mass of the body, which varies with age, sex and existing pathology. Body cell mass (BCM) is difficult to derive but best correlates with isotope measurements of potassium. In malnutrition there is a loss of protein and BCM with water retention which expands extracellular volume.

Our views of energy supply to patients requiring PN have been much modified by the finding that resting energy expenditure (REE) in uncomplicated convalescents is only 10% greater than the preoperative state. In patients with multiple injuries this increases to 10–30% over two weeks which correlates with a period of nitrogen excretion. The highest levels of nitrogen excretion are often associated with fever. In depleted patients and normal adults, nitrogen balance can be increased by increasing either nitrogen or energy intake but depleted patients only can achieve a positive nitrogen balance at zero energy balance. Where REE can be measured and repletion is required, the patient can be given 1.25–1.75 daily REE for calorie requirements. In both the obese and non-obese measurement of energy in a whole body calorimeters show values well below $1.5 \times$ BMR.

The diet should be as complete as possible bearing in mind existing depletion (Table 12.5) and wherever possible the gastro-intestinal tract should be used for feeding since metabolic effects

Table 12.5 Recommended allowances of nutritional substances.

	Allowances kg^{-1} daily body weight to adults.	
Water	30	ml
Energy	30	kcal = 0.13 MJ
Amino acid nitrogen	90	mg (0.7 g amino acids)
Glucose	2	g
Fat	2	g
Sodium	1–1.4	mmol
Potassium	0.7–0.9	mmol
Calcium	0.11	mmol
Magnesium	0.04	mmol
Iron	1	μmol
Manganese	0.6	μmol
Zinc	0.3	μmol
Copper	0.07	μmol
Chlorine	1.3–1.9	mmol
Phosphorus	0.15	mmol
Fluorine	0.7	μmol
Iodine	0.015	μmol
Thiamine	0.02	mg
Riboflavine	0.03	mg
Nicotinamide	0.2	mg
Pyridoxine	0.03	mg
Folic acid	3	μg
Cyanocobalamin	0.03	μg
Pantothenic acid	0.2	mg
Biotin	5	μg
Ascorbic acid	0.5	mg
Retinol	10	μg
Ergocalciferol or cholecalciferol	0.04	mg
Phytylmenaquinone	2	μg
Alpha-Tocopherol	1.5	mg

Table 12.6 Characteristics of some of the available enteral feeds.

Daily Products	Package size	Amount supply approx 2000 kcal	Dilution needed	Lactose free	Problems Features
Clinifeed Iso* (Roussel)	375 ml can	×6	No	No	Low sodium
Clinifeed 400* (Roussel) (also available Clinifeed Protein Rich Vanilla)	375 ml can	×5	Yes	No	
Clinifeed Flavour (Roussel)	375 ml can	×6	No	Yes	
Trisorbon MCl BDH	85 g sachet	×5	Yes	Watch contains no carragenan	Minimal sodium
Ensure* (Abbott) (Also available Ensure Plus)	240 ml cans or bottles	×8	No	Yes	
Isocal* (Mead Johnson)	250 ml can	×8	No	Yes	
Nutrauxil* (Kabi-Vitrum) (Nutrauxil Sip Feed available too)	500 ml bottles	×4	No	Trace	
Flexical* (Mead Johnson)	545 g can	×1	Yes	Yes	Elemental
Vivonex* Eaton Labs) (Also available Vivonex HN)	80 g sachet	×6	Yes	Yes	Elemental preparation
Fortison Standard (Cow & Gate) Also available 'Energy Plus' 'Low Sodium' versions Also available 'Fortisip', 'Fortimel', suitable for 'sip' drinks	500 ml	×4	No	Trace	
Fortison Soya	500 ml	×4	No	Trace	

Flavoured/ Unflavoured	Protein (g)	Fat (g)	CHO (g)	Calories	Osmolarity (mosmol litre $^{-1}$)	Price per 2000 kcal
Vanilla	63	92.4	294	2250	270	£6.00
Vanilla	75	67	275	2000	306 (dil.) 307 (undil.)	£4.75
Neutral or coffee	85	74	315	2250	365	£5.10
No	81	81	240	2000	238	£5.00
Vanilla	70	70	275	1930	380	£7.90 (Bottles) £7.30 (tins)
No	64.1	83.2	252	2008	300 (osmolality)	£4.80
Vanilla	76	68	276	2000	350 (osmolality)	£4.60
Unflavoured & orange vanilla & fruit punch	45	68	308	2000	500 (osmolality) normal dil.	£9.36
No	38	2.6	4.8	1800	500	£7.82
No	80	80	240	2000	260	£4.80
No	80	80	240	2000	260	£5.00

of nutrients given by this route are probably better than by the intravenous route, visceral protein synthesis may improve, and it is associated with far fewer complications than the intravenous route.

Enteral feeding

Characteristics of the currently available feeds are shown in Table 12.6.

Liquid diets prepared in hospitals readily become contaminated with yeasts and staphylococci and are inclined to block fine bore nasogastric tubes. Commercially available liquid diets are safer and easier to handle. Palatability is important where these feeds are used without a nasogastric tube. Administration is ideally by fine bore nasogastric tube but these carry a risk of displacement and pulmonary aspiration of feeds. The presence of an endotracheal tube, impaired cough reflex and the impossibility of retrograde withdrawal of feed combined with the ready acceptability of these tubes predispose to such pulmonary aspiration. Where gastric emptying is impaired but the intestine is otherwise functional, jejunostomy is invaluable. It is a very simple procedure and either wide or fine bore tubes can be used.

There is very little to choose between the many commercially available feeds. The author prefers Nutrauxil which is a whole protein feed. Two litres (4 bottles) provide 2000 kcal and 12 g N_2. The calorie: N_2 ratio is 143:1 and additional carbohydrate can be added. Two litres contain 68 g fat as sunflower oil, 15% of which is essential FA. The Zn content has recently been increased to good effect but some electrolytes and trace elements are still rather low. Long term feeding requires supplementation with vitamin D. This product is gluten and lactose free with an osmolality of 300 mosmol kg^{-1}.

Diarrhoea is uncommon and most frequently related to antibiotic administration or uneven administration of nutrients. Lactose intolerance as a cause has been overstated and diarrhoea is often blamed on hyperosmolality of feeds. However, in healthy people postprandial hyperosmolality and hypernatraemia is normal, hence the recommendation for measuring osmolality in the fasting state. It is difficult, therefore, to see why hyperosmolality of enteral feeds results in diarrhoea.

The vitamin K content of Nutrauxil is rather high. (1.6 mg daily) and the cost £3.00 per g nitrogen.

The requirements for warfarin have been documented to be raised with use of Isocal due to its vitamin K content. High nitrogen Vivonex results in a rise in blood urea nitrogen compared to solid food and predigested protein in patients with malabsorption.

Pump assisted enteral feeding to ensure a better regulated delivery of the feed is indicated in a few patients with persistent diarrhoea and may obviate the need for PN. Whilst continuous feeding may be more comfortable for the patient and easier for nursing staff it is metabolically less efficient than bolus feeding. Resting oxygen consumption is higher and N_2 retention lower with continuous compared to bolus feeding. An undiluted hypertonic feed gives rise to better nitrogen balance for similar side effects compared to an isotonic feed.

The trace element content of commercial enteral feeds may show a large discrepancy between levels stated by the manufacturers and those actually measured by analysis.

PARENTERAL NUTRITION

Despite the limited evidence of reduced mortality and morbidity parenteral nutritional support is widely used for patients in intensive therapy units. The aim is to give all nutrients required by the body in the appropriate proportion. A major dilemma at present is what is the most suitable energy substrate for which patient; the glucose versus fat controversy. The value of protein sparing regimes in the postoperative period or specific AA therapy in various disease processes is unconfirmed.

Minimal technical detail is given here except where it is relevant to the complications.

Disturbances of water and electrolytes and severe hypoalbuminaemia should be corrected prior to starting PN. Analgesia should be adequate and where it is considered appropriate the stress response obtunded. Peripheral vein infusion is possible if isotonic nutrients are used but usually this method cannot provide for the requirements of the critically ill. The author prefers a central line placed by the infraclavicular approach to the subclavian vein although internal jugular vein cannulation is associated with the highest rate of correct placement. A method of cannulation of the inferior vena cava via the saphenous system is described which involves tunnelling to midthigh level. Whichever route is used, the catheter should be inert and flexible, the best currently available being silicone coated. Administration of fluid requires either a volumetric infusion pump or less reliably a drop-rate counter. Volumetric infusion pumps, flow controllers or proportionating valves help to ensure continuous regulated flow rates.

Glucose or fat as energy substrate

Glucose is the most advantageous sugar for PN although others may induce less hyperglycaemia. Most of the early work on the use of glucose as a calorie source was undertaken in the United States where at the time all fat emulsions were banned. Normal growth and nutrition can, however, be maintained with glucose as the major calorie source. In Europe, Intralipid was successfully introduced as an energy substrate at an early stage. There are, however, still those who feel that Intralipid is useful solely to prevent fatty acid deficiency and to replete fat stores although this is best done by glucose. The main objections to the use of fat in the critically ill arose from the evidence of impaired fat utilization in this group of patients. The raised insulin levels suppress fat mobilization. Some studies in humans, however, show increased fat clearance especially in hypermetabolic patients and these patients can oxidize fat in spite of i.v. glucose administration.

Nitrogen balance is influenced by the amount of N_2 in the diet, the metabolic rate and the quantity and source of non-protein energy. Glucose and Intralipid are probably equally effective in maintaining positive nitrogen balance. The relative advantages of each of these substrates depends in part upon the patients. In depleted, non-catabolic patients, where protein intake is low or absent, fat administration has no effect on nitrogen excretion but carbohydrate reduces nitrogen loss. 100 g carbohydrate is sufficient to replace gluconeogenesis from endogenous protein. When diet is adequate isocaloric amounts of fat or glucose have equal nitrogen retention ability although adaption to the utilization of fat takes several days. In catabolic patients carbohydrate is the main protein sparing substrate although 80% of energy requirements are derived from endogenous fat. Ketoadaptation does not develop and endogenous protein continues to be broken down for gluconeogenesis. Under these circumstances infused fat spares endogenous fat stores but not muscle protein. Addition of insulin to the glucose infusion improves protein sparing and N_2 balance although hypoalbuminaemia may result. A further advantage is that insulin enhances sodium pump activity.

Administration of glucose in excess of an optimal 4 mg $kg^{-1} min^{-1}$, however, is harmful as it leads to increased CO_2 production and fat deposition in the liver. Energy cannot be lost from a biological system so that where carbohydrate is given in excess of requirements, about 20% increases REE and 80% is converted to fat with respiratory quotients (RQ) increasing to 7–9. This is associated with an increase in noradrenaline suggesting that the increase in sympathetic response is the cause of the increase REE. Sympathetic induced thermogenesis also occurs constituting a further stress to the hypercatabolic injured patient. Increased CO_2 production leads to respiratory distress or failure in patients with compromised lung function. Oxygen consumption rises particularly in hypercatabolic patients and waste oxidation of FA occurs.

On balance a mixed energy source which combines the advantages of each and obtunds the disadvantages is recommended. An energy source of half Intralipid, half glucose is suitable for the majority of patients requiring PN. Catabolic patients require at least 60% glucose. The hormonal profile of insulin suppression and elevation of the counteracting hormones glucagon, catecholamines and cortisol forms a strong theoretical basis for use of fat emulsion in the seriously ill population.

Glucose

Administration of the highly concentrated solutions required for PN results in local venous complications and systemic problems due to hyperglycaemia and the hypertonic solution.

Fat

Intralipid is the most widely used fat source in this country. The particle size of this emulsion is in the same range as chylomicrons (CM) but its structure differs as it contains no apoproteins or cholesterol. It is probably metabolized similarly to CM by lipoprotein lipase and is removed by the reticuloendothelial system (RES).

Lipid emulsions in some adults and neonates can reduce PaO_2 and diffusing capacity. Pulmonary fat has been shown to accumulate in the lungs of preterm infants fed Intralipid in less than the recommended maximum dose. In injured adult patients an alternative Lipofundin did not produce any change in oxygenation. Intralipid may also enhance the risk of bacterial sepsis. The serum of some acutely ill patients agglutinates Intralipid. This reaction is thought to be due to C reactive protein in the presence of calcium ions. Such patients show evidence of microemboli which could have been due to this agglutination at postmortem. Intralipid should not therefore be given to such acutely ill patients. In rabbits with oleic acid damaged lungs Intralipid infusion increased pulmonary production of vasodilating prostaglandins and hypoxaemia.

Critically ill patients may have reduced intracellular carnitine content which might impair oxidation of long-chain FA. Further investigation is required to determine whether carnitine should be added to the regime or medium chain FA substituted.

MAINTAINENCE OF NITROGEN BALANCE

Protein sparing

The effect of AA solutions after elective surgery has been found to prevent the malnutrition that develops in control patients. This effect is due to AA themselves and not to avoidance of glucose. Some authorities recommend use of AA solutions in those patients who require nutritional support but may be expected to return to oral intake after a few days or in traumatized patients who may fail to absorb from the gastrointestinal tract for some time. Others believe this to be an expensive way of giving water or that this N_2 sparing is no better than infusing an isocalorie amount of carbohydrate. Where nutrient intake must be limited 10 g N_2 and 1000 kcal total energy intake daily may provide optimal sparing of body cell mass.

Amino acid solutions

PN regimes require crystalline 1-AA solutions as a nitrogen source. There is considerable controversy regarding the value of specific AAs in the synthetic crystalline AA mixtures available commercially, in particular those which contain high concentrations of glycine. Glycine may be catabolized to free ammonia rather than contributing useful N_2 for transamination reactions. The limits of glycine turnover of 200 mg kg^{-1} daily are easily exceeded in some high N_2 solutions available. Drug & Therapeutics Bulletin suggests the recosting of AA solutions excluding glycine. These solutions must contain sufficient concentrations of all essential AA. The optimal non-essential AA content of the diet is unknown but egg protein is taken as the standard (51% essential AA).

Very few patients can tolerate more than 20 g i.v. N_2 daily since such levels saturate the hepatic metabolic pathways resulting in deamination of AAs and increased urea production. AA infusion increases minute ventilation, oxygen consumption and the response to hypoxia and hypercarbia. Increased ventilatory drive due to increased protein intake may result in dyspnoea with increased rate and work of breathing.

Liver disease

In patients with cirrhosis, infusion of branched chain amino acids (BCAA) is beneficial, inducing a more positive N_2 balance in the postoperative period than a complete AA mixture. Patients with alcoholic hepatitis given 70–85 g BCAA i.v. daily had an improvement in ascites, encephalopathy, plasma bilirubin and albumin compared to controls who were not given this supplement. Where this therapy was continued for 4 weeks mortality rate fell. In liver disease aromatic AAs are not effectively metabolized and act as precursors for false neurotransmitters such as octopamine. BCAAs are low. In encephalopathic patients therapy to reduce gut uptake of glutamine or its conversion to ammonia may be useful.

Renal disease

Deterioration or impending renal failure should not be a reason for withholding PN. Instead early dialysis should be used to control the metabolic state and allow room for feeding.

Nitrogen balance

Several means of improving N_2 balance have been attempted. Remobilization is the best of these. However, malnutrition produces a rise in intracellular calcium and exercise, including physiotherapy, may further increase intracellular calcium and produce more ultrastructural damage unless accompanied by adequate nutritional repletion.

Leucine has some stimulatory effect on protein synthesis *in vitro* and BCAAs were found to increase N_2 retention in postoperative and multiple trauma patients. Improved N_2 balance occurs in severely catabolic intensive care patients with use of BCAA. Some visceral proteins improved and insulin requirements fell. A metabolite of leucine, alpha-ketoisocaproate, can reduce negative N_2 balance and 3-methylhistidine excretion, although the usefulness of this latter parameter has been questioned. This nitrogen sparing may be related to the increased ketosis since ketone bodies inhibit oxidation of BCAA in muscle and their concentration is increased, or to decreased protein degradation since plasma pre-albumin and retinol binding protein were lower or to an effect on liver protein turnover. The effect is unlikely to be due to a change in carbohydrate metabolism.

Prostaglandins are concerned with intracellular protein metabolism; it may therefore prove possible to reduce muscle protein breakdown with inhibitors of prostaglandin synthesis such as indomethacin. Naftidrofuryl, anabolic steroids, somatomedin and possibly proteinase inhibitors may also find a role in improving N_2 balance.

COMPLICATIONS OF
TOTAL PARENTERAL NUTRITION

Infection is a serious hazard of PN especially in the presence of invasive catheters, steroids, and antibiotic resistant opportunistic

Table 12.7 Complications of PN.

Infective
Metabolic and biochemical
Deficiencies
Disorders or water, sodium and acid base
Jaundice, hypoalbuminaemia
Technical complications

organisms and ranges from infusion phlebitis to suppurative mediastinitis and septicaemia. There is some evidence that surveillance skin cultures can identify those patients at high risk of infection although this is controversial and stricter criteria for culture are required. The distinction between true infection and contamination is difficult. Catheter associated infection can be reduced by the introduction of a 'control of infection team' giving proper education, advice and care and filtration of the fluids may also reduce infection. Tunnelling can reduce sepsis where nursing care is less than optimal although these findings have been criticized on the basis of diagnosis of sepsis. Indwelling intravascular catheters such as the Hickman readily become colonized with organisms giving rise to a positive blood culture with gram-positive organisms. This is especially so in the immunocompromised patient and whilst these organisms give rise to fever they may produce no other clinical sequelae. Further confirmation and evaluation of this situation is required.

Hyperglycaemia requires insulin therapy ideally by continuous infusion using a dynamic scale. Insulin requirements vary throughout the day; the 'dawn phenomenon' of requirements increasing towards dawn is likely to be due to a surge of growth hormone at that time. Haemodynamic effects of infusion of hyperosmolar glucose include expansion of the blood volume, increase in stroke volume and reduction in pulmonary vascular resistance and wedge pressure.

Metabolic bone disease. Severe bone pain may occur with marked disability in the presence of normal Ca, PO_4, 25-$(OH)D_3$ and parathyroid hormone. Bone biopsy may show osteomalacia and hypercalciuria can occur both of which resolve on discontinuing PN. Hypercalcaemia may be precipitated by oliguria. The possibility remains that this is due to administration of excessive phosphate.

Deficiencies

Deficiency of any and every dietary component has been described; only a few important ones are included here.

Zinc deficiency may be overt or subclinical and is due in part to the low level of zinc in some TPN solutions and in part to the formation of Zn-AA complexes with histidine and cysteine which are excreted in the urine. The effects are delayed wound healing and susceptibility to infection.

Phosphate depletion is common in postoperative patients especially when little blood has been transfused with phosphate in the anticoagulant solution. Iatrogenic hyperphosphataemia may occur accompanied by hypocalcaemia and hypomagnesaemia hence the great importance of biochemical monitoring. Deficiency of magnesium alone is not uncommon.

Magesium deficiency. Even with Mg supplements certain groups of patients are susceptible to overt Mg deficiency especially patients with inflammatory bowel disease. 5–10 mmol are required i.v. daily to prevent this or 60 mmol if the oral route is used.

Unless Intralipid is included in the regime, essential *FA deficiency* is likely. Linoleic acid cannot be synthesized and it is therefore recommended that 5–10% of calorie intake should be in the form of this essential FA. Deficiency cannot be prevented by topical application of corn oil. Cases of acute polymyopathy associated with PN have been described which respond to discontinuation of the feeding regime or i.v. lipid supplementation. In these cases the aetiology was thought to be essential FA deficiency. Recent evidence has shown beneficial effects of polyunsaturated fatty acids in fish. A diet rich in these FAs increased eicosapentaenoic acid content of neutrophils and monocytes and inhibited 5-lipo oxygenase pathways of arachidonic acid metabolism and leukotriene B_4 mediated inflammatory reactions. Dietary fish oils rich in omega -3 FAs reduced plasma lipid levels in normal patients and in those with hyperglyceridaemia. It seems likely that modification of fat supply will take place in future.

Intrahepatic cholestasis is not uncommon and may be related to intestinal overgrowth of anaerobic bacteria. The raised serum enzyme levels can be prevented by metronidazole.

Thrombosis and embolism

Subclavian vein thrombosis occurs in an estimated 5–35% of catheters used for PN. Heparin and filtration may reduce the incidence of phlebitis in peripheral infusions. Inclusion of heparin in the PN solution is therefore logical. In patients receiving PN, levels of antithrombin III are reduced. Intralipid has been incriminated in blockage of Hickman and Broviac catheters when mixed with all other nutrients for a prolonged period. This remains to be confirmed.

Catheter embolus is a serious complication which can also occur with tunnelled lines. Major complications such as cardiac perforation, pulmonary thrombosis and dysrhythmia are likely if catheter emboli are not removed. Transvenous, nonsurgical retrieval techniques are described. Paradoxical air embolus has also been described associated with a cracked filter attached to the central PN feeding catheter. Cardiac tamponade due to the central catheter is a potentially fatal complication.

Air embolus can be prevented by use of certain i.v. filters although these do reduce the flow rate and cannot be used with fat emulsions.

PN in respiratory failure

The effects of PN on oxygen consumption and CO_2 production have already been discussed. Diaphragmatic muscle fatigue occurs in malnutrition. The aim of feeding in respiratory failure is to improve respiratory muscle function and sensitivity of hypoxic drive and PN may allow earlier weaning. Ventilator dependent patients who respond to nutritional support by increasing protein synthesis are more likely to wean from mechanical ventilation than those who do not.

PN and drug administration

Antitumour drugs may have an effect on specific nutrients. 5-fluorouracil, perhaps in combination with other drugs may produce thiamine deficiency. There is some evidence that this drug is better tolerated in patients receiving PN.

The effect of PN regimes on oxidative drug metabolism has recently been highlighted. It is known that diet can markedly influence drug metabolism. In a study in volunteers a change from i.v. dextrose to AAs resulted in an increase in antipyrine metabolism. Patients receiving PN may have a variety of other disturbances of organ blood flow and drug interaction to complicate the issue. However it is wise to be cautious with administration of drugs to patients receiving PN.

PLANNING THE REGIME

There is little to choose between currently available PN solutions and it is not the intention to specify one regime here. Far better that the prescriber becomes familiar with a few regimens and understands the principles involved.

Although some centres prefer a standard feeding prescription most patients and especially the critically ill benefit from individual planning. The factors determining design of an optimal regime are considered elsewhere. Although carbohydrate infusion has a progressive N_2 sparing effect, and N_2 balance is directly related to calorie intake, the increased negative N_2 balance associated with severe surgical stress cannot be completely prevented.

The ideal non-protein to calorie ratio varies from 300:1 in starvation to 150:1 in hypercatabolic patients. (Table 12.8).

Table 12.8 Optimal calorie to nitrogen requirements.

	Starving	Catabolic	Hypercatabolic
N_2(g/24h) requirements for equilibrium	7.5	14	25
Kcal (total including protein)	2000	3000	4000
Non-protein calorie to N_2 ratio	250	200	135

Where fluid intake is restricted a reasonable approach is to give as much protein as possible with a calorie to N_2 ratio as low as possible with a mixed fuel energy supply since glucose alone is inclined to lead to more fluid retention than other energy substrates. Insulin is given by a separate infusion at a rate determined by regular blood glucose estimation.

Ethanol is no longer included in PN regimes owing to adverse metabolic effects which include hypoglycaemia, increased blood lactate, 3-hydroxybutyrate and FFAs, reduced growth hormone despite hypoglycaemia and raised cortisol.

The recent introduction of a three litre bag for infusion of a mixture of 24 hour nutritional requirement has proved popular. There seems to be no deterioration of AA and glucose in this system nor changes in concentration of major electrolytes. Vitamins and trace elements were less stable and it is no longer recommended that low concentrations of vitamin C such as contained in Solivito be added to a 3 litre bag as all would be lost

Table 12.9 The composition of one ampoule of Solivito, Vitilipid and Addamel.

Product	Ingredient	Quantity
Solvito	Vitamin B_1	1.2 mg
	Vitamin B_2	1.8 mg
	Nicotinamide	10 mg
	Vitamin B_6	2.0 mg
	Pantothenic acid	10 mg
	Biotin	0.3 mg
	Folic acid	0.2 mg
	Vitamin B_{12}	2 µg
	Vitamin C	30 mg
Vitlipid	Retinol	75 µg (250 iµ)
	Calciferol	0.3 µg (12 iµ)
	Phytomenadione	15 µg
	Fract. soybean oil	100 mg
	Fract. egg phospholipids	12 mg
	Glycerol	25 mg
	NaOH to pH	8
Addamel	Calcium	5 mmol
	Magnesium	1.5 mmol
	Ferric iron	50 µmol
	Zinc	20 µmol
	Manganese	40 µmol
	Copper	5 µmol
	Fluoride	50 µmol
	Iodide	1 µmol
	Chloride	13.3 mmol

by degradation within 3 hours. Calcium may be precipitated if magnesium is low or pH high. If Intralipid is included in these mixtures it causes a significant reduction in drop size of up to 40% depending upon the concentration of the Intralipid or divalent cations, the amino acid solution and the presence of certain vitamin additives. The stability of parentrovite is limited in the presence of light to six hours.

The content of vitamin and trace element additives in common use is given in Table 12.7.

If surgery is required in a patient receiving PN great care must be taken with the lines, to reduce the risk of sepsis. Different i.v. lines should be established for the perioperative period. The importance of perioperative maintainence of glucose homeostasis cannot be overestimated.

Home parenteral nutrition is commonplace in the US and practised by several centres in this country. A register of cases has been set up and the service is likely to develop in a similar way to home dialysis. A dedicated unit with strict protocols is essential for success.

As malnutrition is corrected by feeding the rate of restoration of body cell mass falls to zero at normal nutritional state. At this point N_2 balance will never exceed zero unless the individual is 'body building'. Planning individual PN regimes and their cost-effectiveness may be aided by the use of computers.

Rational prescription of PN in the critically ill must take into account multisystem dysfunction encountered in these circumstances. Knowledge of metabolic derangements in severe illness and attention to detail are outstandingly important.

Chapter 13
Problems of Renal Disease

The human body contains more nephrons than are required to maintain external balance of water and electrolytes but there is progressive reduction in renal blood flow, glomerular filtration rate and increasing glomerular sclerosis after the third decade of life. Excess protein in the diet increases renal blood flow, GFR and therefore 'reserve' glomeruli are used. Diabetes, acquired renal disease and surgical loss of renal mass accelerate development of glomerular sclerosis leading to more rapid loss of renal function. This suggests control of protein intake in the early stage

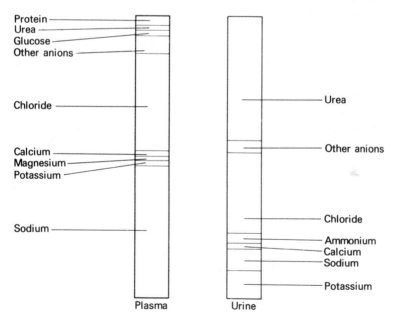

Fig. 13.1 Constituents of plasma and urine. It must be remembered that wide variations will occur in urine constituents during illness.

of renal disease and rigid control of blood sugar by insulin in diabetes. Captopril and some other anti-hypertensive drugs increase creatinine as arterial pressure falls.

As renal disease produces progressive loss of renal substance or an acute event precipitates renal failure profound effects on fluid and electrolyte balance are to be expected. Today sophisticated management in specialist centres can significantly reduce mortality and morbidity in renal disease. Figure 13.1 shows the expected plasma and urine constituents in health.

ACUTE RENAL FAILURE

This is still a devastating illness occurring uncommonly (1–2 per thousand acute hospital admissions), which is often diagnosed by the sudden onset of oliguria. The definition of oliguria in these circumstances is less than 400 ml of urine over 24 hours in an adult. Early diagnosis and referral are essential since the mortality of acute renal failure (ARF) in a nonspecialist unit is 90%.

Clinical accompaniments of ARF

1 Blood urea, creatinine and uric acid are raised.
2 H, K, Mg, PO_4 and sulphate levels in plasma are increased.
3 Impaired salt and water excretion leads in severe cases to oedema. Water retention is usually in excess of salt retention.
4 Plasma HCO_3 falls because HCO_3 buffers the acidic ions which the kidney is unable to excrete. Hyperventilation usually occurs but fails to completely prevent the fall in pH.

Most patients will be oliguric but the same clinical picture may occur in the presence of polyuric renal failure, when a large urine volume of low osmolality is produced.

Causes of acute renal failure

These classically can be usefully divided into prerenal, renal and postrenal causes.

Prerenal causes

1 *Hypovolaemia*:
   haemorrhage
   burns
   other large plasma losses
   dehydration.
2 *Cardiogenic shock*:
   postmyocardial infarction
   massive pulmonary embolus
   aortic dissection
   following cardiopulmonary bypass.
3 *Septicaemia*
4 *Liver disease*:
   cirrhosis
   jaundice
5 *Acute haemolysis*:
   incompatible blood transfusion
   falciparum malaria.
6 *Obstetric disasters*:
   eclampsia
   septic abortion
   postpartum haemorrhage.
7 *Acute pancreatitis*.

Malignant hypertension, hypercalcaemia, drug overdose, hyperuricaemia and renal transplantation may also be aetiological factors. Renal failure associated with non-steroidal anti-inflammatory drugs (NAID) is being reported with increasing frequency. It may present as acute renal failure, tubular necrosis or acute interstitial nephritis. The mechanism is prostaglandin antagonism. These drugs antagonize the diuretic action of frusemide. Chronic renal failure may be the first presentation. Aspirin inhibits formation of the potent vasodilator prostaglandin E reducing renal blood flow and producing medullary ischaemia especially in patients with volume or salt depletion, cardiac failure, chronic renal failure or cirrhosis. Drugs such as beta-blockers that otherwise have no effect on renal function may reduce GFR by reducing cardiac output. Circulating adrenaline increases in shock and indirectly stimulates renal salt conservation. Use of captopril in renal artery stenosis, small vessel disease or after transplantation can precipitate renal failure. It seems likely that inhibition of angiotensin II with reduced GFR is to blame especially as the effect is enhanced by combination with diuretics and suggests that an intact renin-angiotensin system is crucial to autoregulation of GFR at low renal perfusion pressure. It is recommended that radioisotope venography is performed after beginning captopril treatment in patients who have renal artery stenosis and those in whom captopril is being used as the third drug for hypertension where renal artery stenosis has not been excluded.

A prerenal cause for renal failure may be obvious. The two most important aetiological factors are acute circulatory failure and a disturbance of electrolyte balance. In cirrhosis endotoxin may cause hepatorenal syndrome. Recognition of situations in which ARF is likely have lead to improved prophylaxis.

If the patient's condition permits, blood pressure measurements in the supine and upright positions may demonstrate a postural fall in blood pressure indicating a reduced blood volume. Initially simple measures such as urine specific gravity, microscopy and examination of a spun urine specimen may be helpful.

Early appraisal of plasma volume, cardiac function, central venous pressure monitoring, an indwelling arterial cannula for pressure monitoring and Swan Ganz catheter for pulmonary capillary wedge pressure monitoring may provide information for therapy to prevent ARF. It is important also to recognize that severe infection in the surgical patient is very likely to be associated with ARF.

If a patient is known to be at high risk during surgery, for example those with obstructive jaundice, prophylactic treatment in this case with mannitol may avert renal failure. Patients for lower bowel surgery should receive prophylactic antibiotics. In the absence of pre-existing renal disease patients with prerenal ARF usually have a disproportionate rise in blood urea compared to creatinine, and a plasma creatinine greater than 250 mmol l^{-1} is associated with a 90% chance of preexisting renal impairment.

RENAL RESPONSE TO A POOR PERFUSION STATE

Small volumes of urine will be excreted with the following characteristics:

1 A high osmolality, certainly greater than 600 mosmol kg^{-1} and a urine:plasma osmolality ratio of 1.3–2.0.

2 A low Na concentration (less than 10 mmol l^{-1}).

3 A high urine to plasma urea and creatinine ratio greater than 10 to 1.

If the underlying aetiological factors in the development of pre-renal ARF are not corrected then established ARF occurs and then restoration of the circulating blood volume will not improve the situation although it is still essential.

RENAL RESPONSE IN ESTABLISHED ACUTE RENAL FAILURE

Small volumes of urine are excreted with the following characteristics:

1 A urine osmolality which is almost isotonic with plasma (approximately 300 mosmol kg^{-1}) and a urine:plasma osmolality ratio of less than 1.2:1.

2 A high concentration of Na. Urine Na will be greater than 20 mmol l^{-1} or more if diuretics are used.

3 A low urine:plasma urea and creatinine ratio (less than 10:1). Plasma urea and creatinine rise progressively.

In the differential diagnosis of ARF patients with a raised plasma urea or creatinine it has been shown by some authors that tests based on urine constituents are imprecise. In these circumstances the fractional excretion of Na is more valuable. This is calculated as follows from the plasma and urine Na and creatinine.

$$\frac{urine\ Na}{plasma\ Na} \text{ divided by } \frac{urine\ creatinine}{plasma\ creatinine} \times 100$$

If this value is greater than 1 then ARF is due to:

1 Acute tubular necrosis-oliguric, or non-oliguric or

2 Urinary tract obstruction.

If this value is less than 1 then the cause is:

1 Prerenal ARF, or

2 Acute glomerulonephritis.

This calculation should not be used in patients with chronic renal disease, interstitial nephropathy, glycosuria or in those receiving diuretics.

Renal causes of ARF

1 *Rapidly progressive glomerulonephritis*:
post streptococcal glomerulonephritis
Goodpasture's syndrome
polyarteritis nodosa
systemic lupus erythematosus.
2 *Acute interstitial nephritis*. This may be caused by methicillin or
other antibiotics, phenytoin and many other drugs.
3 Haemolytic uraemic syndrome.
4 *Nephrotoxic agents*:
  heavy metals
    lead
    gold
  inorganic compounds
    aniline dyes
  organic compounds
    paraquat
    carbon tetrachloride
  radiographic contrast media.
Mercuric chloride used in peritoneal lavage fluid is very dangerous
and its use should be abandoned.

Despite its high metabolic rate, the kidney is relatively rarely
damaged by toxic substances. Carbon tetrachloride and other
solvents can cause renal tubular damage possibly due to a reactive
metabolite. Inhalation of solvents found in glue and cleaning fluids
may give rise to glomerulonephritis. Inhalation of silica dust may
also be nephrotoxic. Triamterene increases urinary sediment
and this drug may be implicated in the aetiology of interstitial
nephritis. This is quite separate from its effect of inducing hyperkalaemia in patients with renal impairment. The aminoglycoside
antibiotics with the exception of netilmicin are nephrotoxic and
produce a thickening of basement membrane. Serum gamma-
glutamyl transpeptidase does not normally pass into the urine and
its presence is an indication of cellular injury to the proximal
tubule and rises with nephrotoxic antibiotics and frusemide alone
or in combination. Sodium depletion enhances the nephrotoxicity of
amphotericin. Renal biopsy confirms the diagnosis and management is essentially that of the underlying condition, stopping all
nephrotoxic agents and treating the associated hypertension.

Acute decompensation of renal function may occur in patients with pre-existing chronic renal failure.

Postrenal causes of ARF

1 Renal calculi.
2 Analgesic nephropathy producing papillary obstruction.
3 Retroperitoneal fibrosis.
4 Stricture.
5 Carcinoma:
   bladder
   cervix
   prostate
   rectum.
6 Dysproteinaemia: multiple myeloma.
7 Crystalluria: sulphonamides.
8 Hyperuricaemia.

Management includes confirmation of the diagnosis by high dose excretion urography, surgical investigation and treatment of the underlying cause.

In ARF renal blood flow is reduced (40% of normal) but GFR is almost zero. Renal renin levels are raised and locally produced angiotensin II in the renal cortex may be the cause of afferent arteriolar vasoconstriction. Intravascular coagulation is known to be important in the production of many cases of established renal failure.

Following ARF due to prerenal causes an oliguric phase lasts 1–3 weeks with a variable increase in plasma urea and creatinine, an increasing acidosis and hyperkalaemia. Urine output subsequently rises during recovery and the diuretic phase begins.

Mannitol test

A response to mannitol distinguishes a hypovolaemic situation from established ARF. 100 ml of 20% mannitol are infused over 10 minutes. Urine output should increase by 50% over the next 2 hours. If this is not the case then the test can be repeated to a maximum of 100 g mannitol over 24 hours. If the blood volume is known to be high then this test may precipitate heart failure and doses of frusemide up to 1 g given slowly should be used instead. The response to a small conventional dose should be tried first.

Normally mannitol infused i.v. cannot cross cell membranes and therefore it mobilizes cell water. In the presence of normal renal function 80% is excreted and the remainder slowly eliminated by metabolism and with bile. Mannitol intoxication occurs in renal failure as it cannot be excreted. This results in sever hyponatraemia, a large osmolar gap and fluid overload requiring haemodialysis.

Dopamine

Endogenous dopamine is formed in the renal tubular cells from circulating L-dopa, where it acts on specific receptors resulting in dilatation of blood vessels and natriuresis by an effect on tubular transport mechanisms. Abnormalities in renal dopamine production occur in hypertension and oedematous disorders. Some workers have used a low dose dopamine infusion in incipient acute tubular necrosis to prevent oliguria. $0.5–3$ µg kg^{-1} min^{-1} dopamine stimulates renal dopamine receptors producing vasodilatation of the afferent arterioles and a diuresis. The action of frusemide may be synergistic because the dopamine induced vasodilatation allows frusemide access to the loops of Henle whence it increases salt and water excretion. Increased urine flow at the macula densa may inhibit the renin-angiotensin system and hence antagonize the vasoconstriction in afferent arterioles induced by angiotensin II.

Frusemide combined with dopamine prevents the inhibition of prostaglandin migration to the macula densa which may be an important factor in early renal failure.

In established ARF administration of high dose dopamine (15 µg kg^{-1} min^{-1}) may have a deleterious effect by redistributing renal blood flow. This is directly analagous to the alpha vasoconstrictor effects of dopamine on the peripheral circulation. Therefore a small dose of dopamine may be used for its direct renal effects. Less than 10 µg kg^{-1} min^{-1} will produce a positive inotropic effect on the myocardium which may improve GFR and urine flow. Doses greater than this are contraindicated since peripheral vasoconstriction will occur and renal cortical necrosis may be precipitated.

Management of acute renal failure

1 Identify all underlying causes. Chronic renal failure may coexist with acute hypovolaemia in large bowel obstruction. Always suspect and look for septicaemia.

2 Fluids. Fluid management follows the lines suggested in Chapter 11 on preoperative fluid resuscitation. The CVP should be kept at 6–8 cm water.

3 Appropriate treatment should be given for cardiogenic shock and may include the use of inotropic agents or sodium nitroprusside.

If oliguria persists after resuscitation the use of diuretics should be considered (see above). The rapid injection of frusemide in high doses may be associated with ventricular arrythmias. Bumetanide may be a better loop diuretic in these circumstances. If oliguria persists despite two large doses of diuretic then ARF is established.

Management of fully established acute renal failure

Assessment of fluid balance

Accurate fluid balance monitoring is essential since overhydration readily produces pulmonary oedema. All fluid lost should be measured and its electrolyte content ascertained in the laboratory. Losses should be replaced.

Daily weighing is invaluable. Weighing beds are available on many renal units. Weight gain is almost always due to fluid retention.

Fluid administration

Fluid should be given to replace:
1 Insensible loss. In an adult this is about 500 ml daily.
2 The previous day's urine output.
3 The previous day's other losses.
Febrile or burned patients will have increased requirements.

Electrolyte administration

SODIUM

Urinary and other measured Na losses should be replaced unless heart failure or oedema dictate Na restriction.

POTASSIUM

Plasma K is likely to be raised. In addition to the renal failure, sepsis and hypercatabolism will increase plasma K by release of intracellular K. If plasma K exceeds 6 mmol l^{-1} urgent measures are required to reduce it and prevent cardiac arrest in asystole.

Methods of reducing plasma potassium acutely
1   Soluble insulin 10–20 units should be given intravenously with 25–50 g glucose. This manoeuvre promotes uptake of K into cells with glucose facilitated by insulin.
2   Calcium chloride or gluconate 5–10 mmol given intravenously will protect the heart against the adverse effects of a raised K. This should only be given with continuous ECG monitoring. The dose may need repeating.
3   Sodium bicarbonate should be given intravenously. This will correct the metabolic acidosis and tend to reduce plasma K. Unfortunately this involves giving a salt load.

Exchange resins. An exchange resin in the calcium form such as calcium resonium 15 g should be given 8 hourly. This is a relatively slow form of treatment. A resistently raised plasma K is an indication that dialysis will be necessary.

A positive nitrogen (N_2) balance will prevent catabolism and it is illogical to starve a patient in ARF. Adequate calories and N_2 must be given to prevent hypercatabolism and cover energy needs whilst dialysis is undertaken to prevent fluid overload and to reduce plasma K, urea and creatinine. 25–30% of lean body mass may be lost during the course of ARF and subsequent weakness will prolong recovery.

Antibiotics may be required for specific infection with modification of the dose for nephrotoxic drugs. Diuretics may also be indicated.

Patients with ARF should be transferred *early* to a specialist unit for further diagnostic procedures and dialysis. A plot of the weight corrected plasma creatinine against time may allow prediction of the time for dialysis. A plasma urea greater than 30 mmol l^{-1} is unlikely to respond to conservative measures.

CHRONIC RENAL FAILURE

Causes of chronic renal failure

1 Glomerulonephritis.
2 Pyelonephritis.
3 Polycystic kidney.
4 Diabetic nephropathy.
5 Hypertensive renal disease.
6 Analgesic nephropathy.
7 Hereditary and other.

Analgesic nephropathy gives rise to renal papillary necrosis and the tubular damage then leads to interstitial nephritis.

As renal function progressively fails there is an increase in blood urea, creatinine and uric acid. When 70–80% of the nephrons are destroyed the kidney is unable to regulate water, salt and acid base balance. One of two things may happen:

(a) Dehydration with reduction in circulating blood volume may occur.

(b) Oedema or overhydration may occur.

If significant tubular damage occurs then the ability to reabsorb water or electrolyte selectively or to produce HCO_3 is impaired and there is a failure to respond to aldosterone and ADH. Plasma HCO_3 falls and the hydrogen ion concentration rises as does that of sulphate, phosphate and other inorganic anions. Anaemia occurs due to marrow suppression because of reduced erythropoietin levels, toxic factors, reduced intake of haematinics, occult blood loss into the gut and defective platelets leading to coagulation problems.

Endocrine disturbance may affect electrolyte balance and secondary hyperparathyroidism and osteomalacia occur.

In a few patients there is an early increase in plasma renin levels and angiotensin induced hypertension occurs. Aldosterone levels may be raised although less often hypoaldosteronism occurs.

Renal osteodystrophy

The kidney normally hydroxylates $25(OH)D_3$ further to the more active metabolite, thus a defect in mineralization in renal failure or after bilateral nephrectomy might be expected. There is some evidence that this second hydroxylation can occur in extra-renal sites. Metabolic bone disease is a problem in chronic renal failure. Osteoporosis, osteonecrosis, osteosclerosis and extra-skeletal calcification can occur. These have become more common with prolonged survival of patients on haemodialysis. Patients on haemodialysis develop hyperparathyroidism which if mild responds to $1,25(OH_2D_3$. Spontaneous remission of secondary

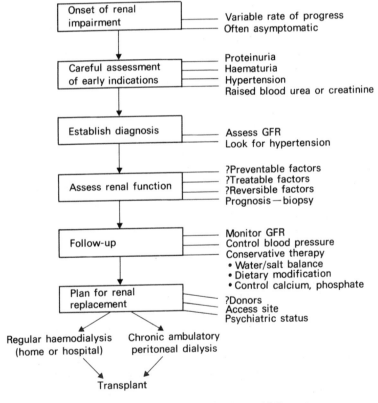

Fig. 13.2 Assessment and management of chronic renal failure.

hyperparathyroidism occurs rarely. Patients subjected to para-thyroidectomy may develop postoperative hypophosphataemia due to reduced mobilization of PO_4 from bone or its increased accumulation in bone. Hypercalcaemic osteomalacia or reduced availability of vitamin D, Ca and PO_4 may be due to aluminium toxicity which also reduces PTH. Aluminium deposition occurs between the thickened osteoid and calcified bone, blocking Ca uptake. Additional Ca may arise from dialysis fluid and vitamin D therapy. These patients have a poor response to $1,25(OH)_2D_3$.

Renal tubular acidosis with uraemia results in rickets in children and osteoporosis in adults. Others develop nephrocalcinosis.

Metabolic bone disease is difficult to manage but desferri-oxamine may be useful for severe dialysis osteomalacia.

Creatinine clearance measurements or serial plasma creatinine may be used to assess progress of chronic renal failure. Clearance is a concept calculated from the concentration of a substance in the urine multiplied by the urine flow per unit of time and divided by the plasma concentration of the substance.
Creatinine clearance =

$$\frac{\text{urine creatinine} \times \text{urine flow per unit time}}{\text{plasma creatinine}}$$

This is usually related to surface area (calculated from weight and height) and the normal range is 40–75 ml min^{-1} m^{-2}.
Very important aspects of management include:
1 Treatment of uncontrolled hypertension.
2 Treatment of urinary tract infection or obstruction.
3 Avoidance of dehydration.
4 Avoidance of nephrotoxic drugs.
In those cases in which water and Na retention occur diuretics may reduce oedema and controlled water and salt intake may reduce plasma creatinine. Plasma K may even fall on dietary restriction if there is impaired tubular reabsorption. Early control of Ca and PO_4 level is important and may delay renal osteodystrophy. A small increase in plasma PO_4 due to reduction in GFR may transiently reduce plasma Ca and stimulate parathormone produc-tion. Phosphate containing enemas may produce serious hypo-calcaemia and hyperphosphataemia in patients with impaired renal function. Bleeding time is altered in most uraemic patients

and they are often anaemic. Platelet mediated haemorrhagic tendency in uraemia can be successfully managed by raising PCV above 30%. The anaemia is usually normochromic and normocytic and does not respond to haematinics. Haemoglobin is about 5–8 g dl^{-1} although it may be higher in patients treated with chronic ambulatory peritoneal dialysis. Several factors contribute to this anaemia including haemolysis and inadequate haemopoietin production.

Methods of reducing plasma phosphate

1 Drugs which bind phosphate in the gut such as aluminium hydroxide will reduce absorption of phosphate but unfortunately aluminium increases the incidence of dialysis dementia.

2 Reduction in dietary phosphate is very difficult but calcium supplements can be given.

3 If bone disease is progressive vitamin D analogues may be indicated (see Chapter 9). Dietary protein restriction is indicated if patients cannot be accepted for dialysis or transplantation. Secondary hyperparathyroidism in children with chronic renal failure can be suppressed by phosphate binders such as calcium carbonate or aluminium hydroxide. These both have the same good effect on bone but the risk of aluminium toxicity suggests that high dose calcium carbonate with dietary phosphate restriction and vitamin D supplements may be the treatment of choice.

Nephrotic syndrome

There is increased permeability of the glomerular basement membrane with leakage of plasma proteins into the urine. As a result plasma protein concentration falls, there is a reduction in circulating blood volume, activation of the renin angiotensin system and aldosterone release. Sodium retention then occurs. ADH secretion is increased and oedema occurs when albumin is <25 g l^{-1}. The degree of oedema depends on the capacity of the liver to synthesize albumin, the age and protein, salt and diuretic intake. The secondary hyperaldosteronism also produces bicarbonate retention, alkalosis and a fall in plasma K. Renal biopsy aids management. If a minimal change appearance is seen these cases respond well to steroid therapy.

Some patients with nephrotic syndrome have high and some low plasma renin activity. In the high renin group this may account for the sodium retention. Although captopril or albumin suppress this to some extent urinary sodium is still less than intake. Therefore, some other mechanism must be responsible and angiotensin converting enzyme inhibitors are unlikely to be of much use therapeutically.

Positively charged particles penetrate the glomerular capillary walls more easily and anionic particles less easily than neutral particles of the same size. There is a fixed negative charge on glomerular capillary walls. Albumin is negatively charged and has a MW less than the limiting pore size of the glomerular capillary wall but because of its negative charge it does not normally leak into the urine. In disease or after i.v. administration of weakly cationized proteins in animals the fixed negative charge may be lost and proteinuria appears.

Difficulties in the diagnosis of renal failure

1 After surgery urinary output is expected to be less than normal so that oliguria as a criterion of renal failure may be misleading at this time (see Chapter 11).

2 Blood urea may rise after surgery due to increased tissue breakdown with fixed free water excretion. Increased water reabsorption will similarly reduce urine output.

3 In an adult a urine flow less than 20 ml hour^{-1} for two consecutive hours is abnormal because it is less than the minimum urine volume found with maximum water conservation and minimum solute load.

Prevention of renal failure at this time hinges on avoiding circulatory insufficiency and maintenance of urine output despite endocrine influences.

Methods of renal protection in the perioperative period

1 Prime the patient with Hartmann's solution to increase ECF by 1 litre. Certain patients are at increased risk of developing renal failure. These include those undergoing cardiopulmonary bypass, aortic clamping, those with obstructive jaundice or suffering severe blood loss. Any technique which reduces sodium reabsorption should be beneficial. This includes administration of mannitol, frusemide or ethacrynic acid or even sodium itself.

2 Mannitol in a dose of 15 g (0.25 g kg^{-1}) may be given over 10 minutes 2 hourly for the duration of the risk to a maximum of 100 g daily. This form of protection is also of value when radiographic contrast media are used in the presence of renal disease. Mannitol, however, causes an increase in cardiac output and pulmonary artery pressure. There is a risk of precipitating heart failure especially in the elderly.

3 Frusemide in a dose of 40 mg 2 hourly. Urine loss should be replaced with 0.45% saline in dextrose with added K if a fall in plasma K occurs. Frusemide and ethacrynic acid increase renal blood flow and GFR during anaesthesia and surgery.

In the situation of extreme fluid loss the oliguria does not always respond to fluid replacement alone but will respond to frusemide or mannitol in addition. If oliguria still persists acute renal failure is established and although this may respond to massive doses of frusemide more usually blood urea and potassium continue to rise.

Renal syndromes associated with non-steroidal anti-inflammatory drugs (NSAIDs)

Some of these have been alluded to already but are referred to again here for the sake of completeness.

1 Acute renal failure — those with pre-existing renal disease or compromised renal perfusion are at greatest risk.

2 Sodium retention and oedema.

3 Hyponatraemia.

4 Hyperkalaemia.

All NSAIDs probably suppress renal prostaglandin synthesis. Cyclo-oxygenase inhibitors may be implicated in the aetiology of papillary necrosis.

5 Nephrotic syndrome.

6 Acute interstitial nephritis.

These drugs should be withheld in patients about to undergo surgery and they should be used with great caution in those at risk; patients with chronic renal insufficiency, congestive cardiac failure, cirrhosis and those with volume contraction of the nephrotic syndrome.

Dialysis

Removal of substances normally excreted by the kidney by dialysis across a semipermeable membrane is now commonplace.

Peritoneal dialysis (PD)

This simple technique can be undertaken in any hospital ward and may be useful for some patients with acute renal failure. It has been used successfully in the management of severe congestive cardiac failure, acute pancreatitis and some cases of poisoning. Recent abdominal surgery precludes this method of dialysis as do extensive burns of the anterior abdominal wall.

The complication rate of this technique is now acceptably low but meticulous attention to detail is essential to reduce infection. A cannula is introduced into the peritoneal cavity avoiding perforation of the viscera, and using a sterile technique. Two litres of dialysis fluid at 37 °C is run into the peritoneal cavity through the cannula and out again into a drainage system. The time and volume of this fluid should be recorded as it is run into the peritoneal cavity and on its return to the drainage system. Accurate weighing of the patient is invaluable during this procedure.

COMPLICATIONS

1 Bowel perforation.
2 Perforation of a blood vessel.
3 Perforation of the bladder.
4 Infection. This is the most important problem. Abdominal wall sepsis may occur or peritonitis due to bacteria or fungi.
5 Failure to drain off dialysis fluid. This is commoner in the obese and may be due to loculation or fibrin obstruction in the catheter. It also occurs if the catheter is misplaced or with an excessively long dwell time. If 2 litres of positive balance occurs PD should be discontinued pending further investigation.
6 Protein loss.
7 Basal pulmonary collapse. This occurs when large volumes of PD fluid splint the diaphragm, reducing movement of the basal lung segments, thus predisposing to collapse.

8 Hypernatraemia.

9 Hyperglycaemia.

10 If fluid is withdrawn from the patient excessively rapidly hypotension and shock may occur.

There are two commercially available PD fluids:

(a) 1.36 g dl^{-1} glucose and osmolality 370 mosmol kg^{-1}. This should remove 1–2 litres of water per day.

(b) 6.36 g dl^{-1} glucose, osmolality 670 mosmol kg^{-1}. This removes more water but is likely to require more insulin for control of blood sugar to prevent hyperglycaemia.

Heparin is sometimes added to PD fluid to help maintain catheter patency. This practice may make traumatic bleeding worse.

The aim in treating ARF is to keep blood urea less than 35 mmol l^{-1} and plasma creatinine less than 1000 mmol^{-1}.

Chronic peritoneal dialysis

This should only be managed in special units. The patient is likely to require about 40 hours each week of peritoneal dialysis. In fact some of the complications of renal failure may improve more readily with PD than with haemodialysis.

Peritoneal dialysis is unsuitable for the management of the hypercatabolic patient as it cannot reduce the blood urea sufficiently. However, it can:

1 Correct acidosis.

2 Return electrolytes to normal.

3 Remove fluid.

Dialysis fluid should always replace essential ions. Mg, Ca, Na and K must be added according to the patient's requirements.

Continuous ambulatory peritoneal dialysis (CAPD)

This is a new technique involving the continuous presence of dialysis fluid in the peritoneal cavity except for periods of 20–40 minutes when the patient changes the fluid for a fesh solution. A soft permanent indwelling catheter is inserted through the abdominal wall into the peritoneal cavity. This is connected via a transfer set to a flexible plastic bag containing the dialysing solution (Fig. 13.3). The bag is raised to shoulder level and emptied into the peritoneal cavity by gravity. The bag is rolled up and retained

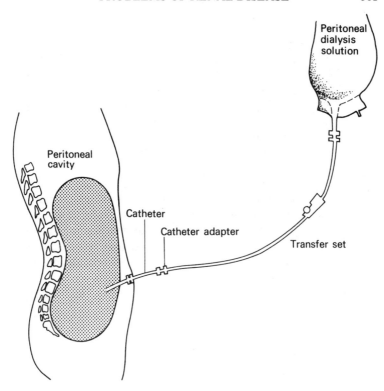

Fig. 13.3 Closed system for continuous ambulatory peritoneal dialysis.

under the patient's clothing during the dwell time and then lowered to below the peritoneal cavity to drain the solution out.

Four to eight hours dwell time is allowed between exchanges. There is a high degree of removal of urea, creatinine and higher molecular weight solutes.

The advantages of the technique are:

1 Better plasma biochemistry with a higher haemoglobin and lower plasma phosphate than with haemodialysis.

2 The cost is about half that of home haemodialysis.

3 Patient independence and rehabilitation. This technique allows a near normal diet, everyday mobility and also allows the renal failure patient to take a holiday.

The main disadvantage is sepsis. Recurrent peritonitis is a problem for which prophylactic cephelexin has proved ineffective and gentamicin accumulates with a very long half life. The patient needs a training period. With a meticulous sterile technique and continuing improvement in catheter designs and connections the incidence of sepsis may be reduced. Sepsis may be reduced by enclosing the catheter connection in a sterile fashion. Early detection of sepsis is important using techniques such as culture of dialysis drainage fluid using the dialysate digest medium tube. This shows a cloudy appearance well before organisms can be cultured so that antibiotics can be given to stop the full blown picture of sepsis.

40% of patients will have one episode of peritonitis in the first year of treatment which occasionally can result in leak of infected fluid per vaginum or multiple hepatic abscesses. Infection is often due to skin organisms. Staphylococcus epidermis accounts for 30–40% of infective episodes, Staph. aureus and gram negative bacilli especially Pseudomonas are also common and fungal infection, usually Candida, less uncommon. No single antibiotic is appropriate for all of these, the duration of treatment is doubtful and prophylactic antibiotics are probably not indicated.

Other complications include Richter's hernia during insertion of the catheter and other abdominal hernias during treatment including hernia of Morgagni. The acetate in CAPD dialysate is associated with abdominal pain and perhaps progressive peritoneal sclerosis.

CAPD is useful in patients over 60 years of age with end stage renal disease who would not fulfil the criteria for chronic haemodialysis. It compares favourably with other techniques with good survival and rehabilitation. It has been used as a home dialysis technique for children with good control of uraemia, nutrition, anaemia, hypertension and mineral metabolism. The cost is relatively low and it avoids fluctuations in control of solute concentration and ECF volume compared to haemodialysis, with increased mobility and independence. In uraemic diabetics insulin can be instilled into the peritoneal cavity, most effectively into an empty cavity at least 30 minutes before the dialysate is instilled. Previous CAPD is not a risk factor for renal transplantation.

CAPD results in lower urea and creatinine clearance than haemodialysis. Some of the toxic metabolites in renal failure are difficult to identify and include some high MW substances which are minimally dialysed.

Haemofiltration

Although haemodialysis can be reduced to 4 hours three times weekly, hypotension, cramp, nausea and headache remain problems and 'middle molecules' between 1–2000 daltons are not well cleared. The technique of haemofiltration which produces an ultrafiltrate of plasma using a highly permeable membrane of up to 6000 daltons has advantages. Blood is withdrawn at a flow rate of about 200 ml min^{-1} and returned with a volume of fluid equal to that withdrawn as required. Blood pressure is more easily controlled with long term use without the occurrence of acute hypotension during filtration which occurs with haemodialysis. Haemofiltration is useful in acute renal failure to make space for nutrition and for treatment of profound diuretic resistant oedema.

Haemodialysis

The great advantage of this technique over PD is its much greater efficiency. Urea clearance is almost four times greater than with PD (150 ml min^{-1} compared to 40 ml min^{-1} for PD). This therefore reduces dialysis time in the chronic situation.

Haemodialysis (HD) allows much greater patient mobility, is much more comfortable for the patient and does not cause protein depletion or basal lung collapse. Although it is more efficient it may be less safe by removing too much fluid from the patient too quickly.

Disadvantages of haemodialysis

1 Special units required.
2 Vascular access may be difficult.
3 Hypotensive patients may have inadequate blood flow through the dialyser.
4 Heparin may be a hazard.
5 Dialysis dementia. The prevalence of this has recently been put at 600 per 100 000 European dialysis patients. Dementia appears to be due to aluminium toxicity.
6 Emotional problems. Dialysis obviously disrupts family life and employment.

7 Dialysis arthropathy is a common potentially severe complication of long term dialysis of unknown cause but amyloid may be implicated.

Vascular access is via:

(a) Arteriovenous fistula.

(b) Arteriovenous shunt.

(c) As a temporary measure percutaneous cannulation of the femoral vein for access to the inferior vena cava.

(d) Increasing use is being made of indwelling subclavian catheters for patients in acute renal failure without a useable arteriovenous shunt. The Tenckhoff catheter is an example. There is the risk of infection so that insertion technique and aftercare must be meticulous and the catheter removed if blood stream infection occurs. The incidence of thrombosis can be reduced by leaving the catheter full of heparin when not in use. Increased ionized Ca occurs during haemodialysis and improves left ventricular contractility during the procedure.

The safe maximum concentration of aluminium in water for use in haemodialysis is 1.0 mmol l^{-1} and 0.5 mmol l^{-1} in CAPD dialysate. Excess aluminium causes dialysis dementia and osteomalacia. It is wise to measure aluminium periodically especially in young children. Aluminium toxicity can occur in infants with uraemia who are not undergoing dialysis but who are being treated with aluminium hydroxide from the first month of life.

WHEN TO DIALYSE

Ideally in ARF blood urea should be kept below 35 mmol l^{-1}. Therefore dialysis should be started early and repeated regularly and frequently. This is of tremendous value in hypercatabolic patients since it permits adequate parenteral nutrition to be administered. In chronic renal failure a creatinine clearance of less than 10 ml min^{-1} should be treated by dialysis. Other indications for dialysis include a very high plasma potassium (greater than 6 mmol l^{-1}) and severe positive fluid balance.

In the best run dialysis units biochemical parameters are not entirely normal. Plasma creatinine is always raised. Severe oliguria usually occurs when haemodialysis is started and fluid restriction may be necessary. Anaemia and renal osteodystrophy may still be problems and there is no doubt that family life is disrupted.

Renal transplantation provides an alternative to chronic haemo-dialysis or continuous ambulatory peritoneal dialysis. After renal allograft hypertension is a significant problem. Its aetiology is multifactorial but the renin-angiotensin system is probably impli-cated. Captopril may be a useful drug in these circumstances provided that there is no renal artery stenosis.

In acute renal failure dialysis should be continued into the diuretic phase since although urine flow increases GFR may recover more slowly. Indications to stop dialysis include an improvement in blood and urine biochemistry.

Daily urea and electrolyte measurements are of course mandatory.

Other therapies that have been tried in renal failure include prostaglandin E1 infusion in patients with chronic glomeru-lonephritis. In some patients renal function improved with a 29% increase in creatinine clearance and sustained benefit for 1–7 months after a course of PGE1.

Chapter 14
Fluid Balance in Special Circumstances

CONGESTIVE CARDIAC FAILURE

Congestive cardiac failure (CCF) may be defined as circulatory insufficiency in which the heart fails to supply the metabolic needs of the body. There is a fall in cardiac output with a reduction in venous return. Organ perfusion is reduced and renal handling of Na and water is inappropriate. There is an increase in plasma volume, ECF volume and ICF volume.

Mechanisms involved in salt and water retention in congestive cardiac failure

1   The reduction in cardiac output reduces GFR, and tubular reabsorption of Na and water are increased. The effective fall in circulating blood volume will increase aldosterone and ADH. The Na and water retention increase the filling pressure of the already

impaired myocardium with further failure. The increased Na retention stimulates ADH secretion via the osmoreceptors.

2 The myocardial failure per se increases venous pressure which increases capillary hydrostatic pressure. This impairs return of fluid into the capillary with development of oedema.

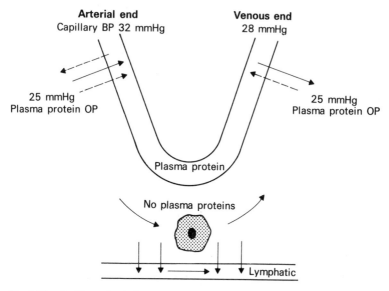

Fig. 14.1 Capillary circulation in congestive cardiac failure.

Urine volume is reduced with an increase in specific gravity and osmolality unless diuretics are administered or there is intrinsic renal disease. CVP and pulmonary capillary wedge pressure are raised. Arterial PaO_2 falls. The acid base status is usually normal but haematocrit and plasma protein levels are reduced by a dilutional effect. Increased circulating levels of vasodilator prostaglandins correlate directly with plasma renin activity and angiotensin II. Sodium concentration is inversely proportional to PGE2 and plasma renin. In severe CCF cardiac noradrenaline is depleted and the number of beta receptors reduced.

Treatment

Treatment includes use of diuretics and digitalis. Sodium and water restriction may be appropriate. In addition nitrates may be used to reduce the preload or nitroprusside to reduce the afterload on the heart.

Hypertension

There are many causes of hypertension some of which also cause electrolyte disturbances. These include the following:

Renal disease.
Endocrine disorders:
   Cushing's syndrome
   Phaeochromocytoma
   Pregnancy
Vascular disease — coarctation of the aorta.
Essential hypertension.
Porphyria.
Acutely increased intracranial pressure.
Lead poisoning.

The electrolyte disturbances are those of the underlying cause but briefly:

1 Hyponatraemia may occur in hypertension treated by diuretic therapy, in renal disease and in accelerated hypertension with secondary hyperaldosteronism.
2 Hypernatraemia occurs in primary hyperaldosteronism.
3 Hypokalaemia occurs in diuretic therapy and increased aldosterone secretion.
4 Hypomagnesaemia occurs in primary hyperaldosteronism.
5 Metabolic alkalosis occurs with thiazide diuretics and in circumstances of excess concentration of aldosterone or corticosteroids. In these situations H and K are exchanged for sodium in the distal tubule with an increase in plasma HCO_3.
6 Total body water is increased in primary hyperaldosteronism and in hypertension with acromegaly. Total body water is reduced in renovascular hypertension.
Measurement of red blood cell mass, plasma volume, ECF volume and total body water in normal people and patients with essential hypertension (and normal renin) shows that in hypertension, plasma volume and total blood volume are lower, there is reduc-

tion in ECF volume but no change in partition of ECF volume between plasma and interstitial compartments.

RESPIRATORY DISEASE

Chronic obstructive pulmonary disease (COPD) associated with hypoxia may be accompanied by secondary polycythaemia and an increase in haematocrit. Plasma exchange in which the patients blood is removed and replaced with dextran or HAS will improve blood flow, viscosity and oxygen delivery to the tissues.

When carbon dioxide (CO_2) retention occurs there is compensatory reabsorption of bicarbonate by the kidney in an attempt to maintain a normal pH.

Dyspnoea and hyperventilation are associated with increased water loss from the lungs and dehydration. Patients requiring controlled mechanical ventilation for respiratory failure show altered renal function especially if positive end expiratory pressure is in use. Urine volume and sodium excretion fall due to decreased venous return and cardiac output. Baroreceptors, renin, ADH and the sympathetic nervous system may also be involved. The role of ADH is largely in defence of the circulating blood volume and changes in free water clearance due to intrarenal mechanisms. These lead to fluid retention including in the pulmonary interstitial space and oedema develops. In this circumstance diuretics may further reduce cardiac output.

For the role of respiratory disease in acid base disturbance see Chapter 10.

GASTROINTESTINAL DISEASE

Pyloric obstruction

In this situation large volumes of gastric juice are vomited. Hydrogen, K, Na and Cl ions are lost with resulting falls in their plasma levels. Fluid loss results in reduction in circulating blood volume. Plasma bicarbonate (HCO_3) is raised and metabolic alkalosis occurs with tetany in extreme cases. Some respiratory depression occurs with retention of CO_2 so that carbonic acid levels rise in an attempt to restore pH to normal. The deficit of K reduces its exchange with sodium and H in the renal tubule so that aciduria may occur. In addition K enters cells in exchange for H. Later starvation with a decreased intake of carbohydrate and protein lead to catabolism with ketosis and an increase in blood urea.

Treatment consists of 0.9% sodium chloride with added KCl. Surgery is the definitive treatment for this condition.

Biliary and pancreatic fistulae

Up to 1 litre day^{-1} of alkaline fluid may be lost. This fluid will have a high Na and HCO_3 content. Metabolic acidosis occurs with a fall in plasma HCO_3, pH and PCO_2 and an increase in Cl to maintain anion balance. Hydrogen concentration is increased and if the situation is prolonged K levels fall.

Diarrhoea

Loss of water may amount to 10 litres daily in severe cases. Na loss may reach 350 mmol daily and K 45 mmol daily. Intestinal juice contains more HCO_3 than Cl so that plasma HCO_3 is more reduced than Cl and metabolic acidosis occurs. Lack of carbohydrate ingestion will lead to ketonaemia. Circulating blood volume will be reduced with oliguria. Exudative diarrhoea will also result in increased protein loss.

Treatment

Fluid loss should be replaced with a solution containing
     0.45% saline + 40 mmol l^{-1} K and 45 mmol l^{-1} HCO_3.
In infants a different formula is used since losses of Na, K and Cl are likely to be less. Mention has already been made of oral replacement solutions for infants with diarrhoea (UNICEF/WHO glucose electrolyte solution). This solution should be used with caution taking into account the particular needs of the locality and age of the patient (see Chapter 7).

Cholera

Vibrio cholerae produces the most dramatic diarrhoeal illness with a mortality in untreated cases of up to 70%. Fluid loss is the primary cause of death. Mild to moderate disease can be treated with oral glucose electrolyte solution but severe cases always need i.v. fluids. Drugs have been tried to reduce the rate of fluid loss. Chlorpromazine seems effective, by virtue of inhibition of adenyl cyclase activity in intestinal mucosa, in reducing the rate of fluid loss, duration of diarrhoea and frequency of vomiting without inducing hypotension provided the patient is adequately hydrated. Chlorpromazine probably produces its anti-secretory effects by an action on calmodulin.

Protein losing enteropathy

Villous adenoma, inflammatory gastrointestinal tract disease, collagen disease and sprue may be associated with daily loss of 40–50 g of protein. This situation leads to a reduction in circulating blood volume and secondary hyperaldosteronism. There will then be a low plasma K and alkalosis. All plasma proteins are low. Treatment is that of the underlying disease, for example, steroids for ulcerative colitis.

Intestinal obstruction

If this is high the losses will be mainly gastric. In lower intestinal obstruction the loss will be as in diarrhoea, but because of the obstructed bowel, inflammation and congestion occur with loss of protein. Treatment consists of replacing the losses, and surgery. Despite marked fluid loss, electrolyte abnormalities are not usually dramatic although oliguria and haemoconcentration occur.

Peritonitis

Sequestration of fluid and electrolytes occurs in the peritoneal cavity. This will be increased by the osmotic effect of exudation of protein. At first there is salt and water loss; later hypovolaemia, hypotension and oliguric renal failure occur.

Malabsorption

Onset is usually gradual with weight loss, hypoalbuminaemia, reduced Ca, Mg, increased alkaline phosphatase, prothrombin time and anaemia. The net effects are a combination of dehydration and starvation with deficiency of fat soluble vitamins A, D, E and K. If malabsorption is severe excessive loss of Na occurs with osmotic diarrhoea, hypokalaemia and metabolic acidosis.

In any of these situations, if due consideration is not given to the nature of the fluid losses inappropriate or inadequate therapy may be given. The electrolyte concentration of secretions is shown in Table A.4

LIVER DISEASE

Acute liver failure

Hyperventilation with respiratory alkalosis occurs but it is unwise to correct this as deterioration in the patients condition may occur. Hypoxia is common and is due to a combination of factors:
1 Central nervous system depression.
2 Pulmonary oedema.
3 Infection.
4 Intrapulmonary shunting.
Hyponatraemia and hypokalaemia occur in 50% of patients at some time during the illness. This is partly dilutional. There is marked renal impairment of free water excretion. Hyponatraemia also occurs due to loss of sodium into the cell when the activity of Na/K ATPase is reduced. This is the sick cell syndrome (see Chapter 7).

Chronic liver disease

Fluid retention occurs resulting in oliguria, ascites and peripheral oedema. In cirrhosis some disturbance exists prior to the onset of ascites in that there is a reduction in total body K with a normal or low plasma urea and inability to handle a water load. Plasma volume and ECF volume are increased with an increase in exchangeable sodium.

As ascites develops plasma protein levels fall especially plasma albumin resulting in a reduction in effective plasma volume and renal plasma flow. This increases activity of the renin angiotensin system with secondary hyperaldosteronism. Aldosterone increases reabsorption of sodium from the distal tubule in exchange for H and K. Therefore there is a temporary increase in plasma Na with reduced K and H and a mild metabolic alkalosis. Water is retained with Na. Cirrhosis results in increased splanchnic capillary pressure which, in conjunction with the overloading of lymphatic reabsorption and reduced plasma colloid osmotic pressure, results in ascites. Although the effective plasma volume is reduced, the actual volume is increased. Some of this is within distended portal veins. Usually the raised plasma ammonia stimulates the respiratory centre and hyperventilation occurs. Hyponatraemia may induce thirst and demeclocycline increases urine flow. Plasma noradrenaline is increased in ascites and other disorders where baroreceptors activate efferent sympathetic discharge. Relocation of ascitic fluid (shunt or immersion) produces a diuresis. It seems therefore that ascites is primarily a renal retention effect and secondary to a reduction in effective blood volume which is worsened by diuretics and may result in the hepatorenal syndrome. Reduction in plasma angiotensin II and aldosterone do not achieve a diuresis, so other non-hormonal factors are likely to be implicated in oedema of cirrhosis. Aldosterone receptors may be hypersensitive in cirrhosis.

Causes of alkalosis in cirrhosis

1 Hypokalaemic alkalosis associated with intracellular acidosis.
2 The respiratory centre stimulation due to hyperammonaemia and hypoxia.
3 Renal excretion of H due to the secondary hyperaldosteronism.

Blood urea is low owing to reduced synthesis by the liver. Urine is often hyperosmolar compared to plasma because the kidney is unable to excrete a dilute urine.

At a later stage of liver failure, renal failure occurs. This is the hepatorenal syndrome. The most likely cause of this is endotoxin production. GFR and renal blood flow may fall rapidly with an increase in plasma renin activity.

Bone disease

Osteoporosis occurs in cirrhosis and osteomalacia in those with cholestatic disease. This is related to low $25(OH)D_3$ and is likely to be due to defective intake, impaired absorption or excessive loss. Reduced exposure to sunlight is important. Parenteral vitamin D is recommended for prevention of osteomalacia in jaundice of greater than one year duration.

Management

1 Diuretics. Potent loop diuretics such as frusemide induce further hypokalaemia and alklosis and therefore should be avoided. They may precipitate hepatic encephalopathy and do nothing to improve renal perfusion. Both thiazides and frusemide produce hyperuricaemia but bumetanide may have an advantage by producing less urate retention.

Hyperaldosteronism is improved by spironolactone. This will conserve K but results in relatively poor excretion of sodium.

Triamterene and amiloride are diuretics which conserve K but there may be progressive resistance to their action if they are used alone.

2 Dietary sodium should be restricted.

3 Paracentesis should only be used to relieve cardiorespiratory distress as it removes a large amount of protein, with a risk of infection. The net result is a further reduction in plasma volume, increase in aldosterone and ADH production.

4 Steroids have been recommended by some workers to increase free water and sodium excretion.

5 Salt poor albumin may be used in an attempt to raise plasma albumin.

ALCOHOLISM

Alcohol is a diuretic which increases free water clearance by inhibition of ADH. This is associated with a mild metabolic acidosis. As Na, K and Cl retention occur with an increase in osmolality, production of ADH is stimulated and an antidiuretic phase then occurs.

Chronic alcoholics have a reduced plasma K. Body sodium is usually increased. It is well recognized that chronic alcoholic beer drinkers may suffer severe hyponatraemia and hypo-osmolality (see Chapter 7).

Malnutrition, vomiting and diarrhoea are common in alcoholics and will have their effects on electrolyte balance. Deficiency of calcium and magnesium occur due to increased renal excretion, malnutrition and increased intestinal loss.

In the normal population ethanol is known to induce hypo-glycaemia. In a recent study infusion of ethanol was shown to produce the following changes:

1 Hypoglycaemia due to inhibition of hepatic gluconeogenesis.
2 An increase in blood lactate, hydroxybutyrate and free fatty acids.
3 Reduced growth hormone concentration.
4 Raised plasma cortisol concentration.

Clearly the inclusion of alcohol in regimes for parenteral nutrition may have serious side effects.

DIABETES

The water, electrolyte and acid base abnormalities in diabetes are due to absolute or relative lack of insulin causing impairment of glucose utilization and gluconeogenesis. In the absence of insulin, glucose uptake into cells is very slow and therefore plasma glucose rises, increasing the effective osmotic pressure of ECF with passage of water from cells to maintain isotonicity.

As blood glucose rises the renal threshold for glucose is reached (10 mmol l^{-1}). Glucose then appears in the urine and the osmotic diuresis prevents the reabsorption of water Na, K and Cl. This state of affairs may continue until considerable dehydration reduces GFR and oliguria occurs despite hyperglycaemia. As dehydration progresses less Na reaches the distal tubule and hence less is available for exchange with K and H and their retention is a significant cause of metabolic acidosis.

The resultant effects of this are to produce progressive cellular dehydration and loss of water together with Na, K and Cl into urine. In controlled hyperglycaemia therefore total body osmolality is raised but plasma sodium and chloride are normal or low because of the dilutional effects of water from cells. If, however, a raised plasma sodium is found this indicates severe water loss (hyperglycaemic hyperosmolar non-ketotic diabetic coma). Direct measurements of plasma osmolality are very helpful and can be used to estimate the water deficit (see Chapter 6). Hyperglycaemia as already discussed is a less potent stimulus for ADH production then hypernatraemia for the same level of osmolality.

In evaluating plasma electrolytes to assess deficits it is essential to ensure that there is no serious hyperlipidaemia producing pseudohyponatraemia which leads to gross underestimation of the extent of the dehydration (see Chapter 7).

Since normal carbohydrate utilization is impaired in diabetes large amounts of fat are metabolized resulting in increased ketone body production. When these organic acids are added to ECF they are initially buffered producing a fall in plasma bicarbonate. Hyperventilation occurs with a fall in plasma $PaCO_2$ in an attempt to compensate for the impending metabolic acidosis. Eventually these compensatory mechanisms become exhausted and pH falls (Chapter 10).

Water loss is in excess of electrolyte loss and glucose is less effective than other solutes in stimulating thirst. Hypertonic dehydration occurs and the patient cannot take sufficient fluids orally because of vomiting, anorexia and impending coma. As the water loss can be extreme a reduction in circulating blood volume, hypotension and shock eventually occur. Abdominal pain may be due to ketosis per se although it is often misdiagnosed as a surgical emergency. A marked leucocytosis may occur. Acute pancreatitis

Table 14.1 Fluids lost in diabetic ketosis

	Osmolality	Sodium concentration
Skin, lungs (water)	0	0
Kidney	Iso or hypertonic	25–100 mmol 1^{-1}
Vomit	Isotonic	50–80 mmol 1^{-1}

may be precipitated. In diabetic coma mortality may be as high as 10% but recent advances in management may reduce this.

Glucose concentrations can be measured by enzymatic means using glucose oxidase or dehydrogenase. Some drugs interfere with such estimations. Portable machines such as the Reflomat reflectance meter utilize glucose oxidase impregnated strips and can be used for repeated capillary blood sugar measurements.

Management of uncontrolled diabetes

The relative amounts of insulin and fluid required depend to some extent on whether diabetic ketoacidosis or hyperosmolar hyperglycaemic non-ketotic coma is present.

Diabetic ketoacidosis

This may be due to a precipitating cause in a stable diabetic, for example, chest infection, myocardial infarction or may occur at presentation in a young diabetic or be precipitated by a stressful event. The diagnosis may be suspected from air hunger (Kussmaul's respiration) and the smell of ketones on the breath and confirmed by urine and blood sugar measurement. Too many patients develop ketoacidosis because of delay in diagnosis or poor management of the diabetic state during intercurrent illness. The more severe the volume depletion on admission the greater will be the ketone retention and the less prominent the hyperchloraemic acidosis. After therapy hyperchloraemia develops in most patients because of retention of Cl in excess of Na as ketones are excreted by the kidney. Fluid loss is mainly due to osmotic diuresis but hyperventilation, pyrexia and vomiting also contribute.

INVESTIGATIONS

Blood glucose.
Urea and electrolytes.
Haemoglobin and haematocrit.
Arterial blood gases.
Blood culture.

These are urgent investigations which should be carried out in the casualty department. Gastric distention is common and a nasogastric tube should be passed.

A large bore cannula sited peripherally is essential. In addition CVP measurement is valuable especially in the elderly in whom large volumes of fluid replacement may result in cardiac failure. One litre of 0.9% saline may be given over 30 minutes whilst results are awaited. 20 units of soluble insulin or the less antigenic neutral or Actrapid insulin for a newly diagnosed patient may be added to the infusion with good mixing. Thereafter 1 litre 0.9% saline may be given hourly or under CVP control. If the patient remains hypotensive with a low CVP after several litres of saline then consider giving HAS or an alternative colloid to increase colloid osmotic pressure.

Increasing use is being made of continuous intravenous insulin infusion with the aim of producing an effective plasma insulin level between 20–200 μ units ml^{-1}. Smaller doses of insulin used in this way are effective (2.4–12 units hourly). This has advantages over subcutaneous injection which is too erratic and over intermittent boluses of intravenous insulin. With these small doses given continuously i.v. blood glucose falls at the same rate as with larger doses of insulin. Hypokalaemia and cerebral oedema are less likely, the increase in blood lactate is reduced and late hypoglycaemia is also less common. Insulin dose will fall dramatically as acidosis improves (0.5–2 units hourly). One method uses a continuous infusion pump to give 6 units soluble insulin hour^{-1}. If the response is unsatisfactory the dose of insulin is doubled, and when blood glucose reaches 10 mmol l^{-1} the dose is reduced. It is very rare to need more than 30 units hour^{-1} unless steroids or sympathomimetics are in use. Increases in insulin within the physiological range stimulate Na reabsorption by the distal nephron independent of changes in other hormones. Insulin produces a fall in plasma K uninfluenced by age or beta-blockade.

Usually blood sugar falls at a rate of about 5 mmol hour^{-1}.

When it has reached 10 mmol l^{-1}, 6 g glucose hour^{-1} should be added to the infusion. Some workers add albumin or haemaccel to the insulin infusion to reduce absorption of insulin to the plastic of the administration system. However, with high concentrations of insulin this is not necessary. Insulin should be continued until all evidence of ketosis disappears.

POTASSIUM

As glucose begins to enter the cells under the influence of insulin it is accompanied by K so that plasma concentration falls and supplements will be required to prevent dangerous hypokalaemia. About 20 mmol K may be given after the first or second litre in each litre 0.9% saline. Some authorities prefer to give K only when plasma K is 4.5 mmol or less when 13–26 mmol hourly are infused. The total dose required may be 50–250 mmol especially if bicarbonate is used. It is mandatory to measure plasma K and glucose hourly to modify subsequent K and insulin dosage and to measure urine ketones and glucose. Less frequent blood gas measurement allows correction of pH.

BICARBONATE

The pH will correct itself over 5–6 hours (after 6–7 litres saline) but if the pH is very low, for example less than 7.1 and has not improved over 2 hours, then NaHCO$_3$ should be used (1.4% NaHCO$_3$, which contains 1 mmol in each 6 ml. 50 ml may be given or 100 ml if the pH is less than 7.0). The danger of this is that a left shift of the O_2 dissociation curve occurs with impaired release of O_2. The low pH in diabetic ketoacidosis results in a low 2,3 DPG which balances the effect of acidosis on the O_2 dissociation curve so that tissue oxygenation is normal. If bicarbonate is transfused rapidly to correct pH, the low 2,3 DPG will be unopposed and the O_2 dissociation curve shifts to the left with a rapid fall in tissue oxygenation. 2,3 DPG levels may take several days to return to normal because phosphate is being rapidly taken up by the cells. Therefore bicarbonate should be used with great care and phosphate added to the i.v. regime when plasma PO$_4$ is less than 0.16 mmol l^{-1}. Infusion of a significant amount of HCO$_3$ (150 mmol) results in a delayed fall in lactate and total ketone bodies compared to regimens using saline only. In an analysis of 95 episodes of severe diabetic ketoacidosis in patients treated by conventional means there was no difference in outcome between those treated with or without HCO_3.

As conscious level improves oral fluid may be given. When the patient is eating, intravenous insulin may be discontinued and subcutaneous insulin give b.d. or t.d.s. The intravenous infusion should be continued for about 48 hours. If the blood glucose fall is dramatic then 5% dextrose may be substituted but with small doses of intravenous insulin this is rarely necessary if the patient is allowed to eat as soon as he is able.

Urine and plasma osmolality are a valuable guide to treatment. Complications of diabetic ketoacidosis include; shock, infection, arterial thrombosis and cerebral oedema especially in children. Subclinical brain swelling diagnosed by CT scan is common in children and adults and occurs after improvement in hyperglycaemia and arterial pH. It may be associated with excessive use of insulin.

Cerebral capillary endothelium, the site of the barrier and transport functions of the blood brain barrier is exposed to abnormal osmotic gradients during treatment. This endothelium is almost impermeable to Na but plays an active part in regulating volume and composition of brain ISF through membrane ATPase. This activity is modified in hyperosmolality in a way which assists in maintaining brain water in dehydrated states. Injury to this endothelium releases substances that rapidly increase brain swelling. This is associated clinically with a fall in colloid osmotic pressure, PCV and arterial PaO_2 potentially due to excessive crystalloid loading. Although shock lung may occur in ketoacidosis, in association with acute disseminated intravascular coagulation, acute respiratory distress is more likely due to a reduction in COP and hypoalbuminaemia caused by high volume crystalloid infusion.

Hyperosmolar hyperglycaemic non-ketotic coma

Hyperosmolar hyperglycaemic non-ketotic coma now forms 10–20% of causes of diabetic coma. This tends to occur in the elderly and usually the hyperglycaemia has been present for weeks before admission but there are no ketones in the urine. Blood glucose can become very high (greater than 40 mmol l^{-1}) and hypertonicity severe (plasma osmolality $>$ 350 mosmol kg^{-1}). Plasma total amino acids are low, possibly due to excessive hepatic extraction. Hyperglycaemia may be precipitated by K depletion as this syndrome can be produced by diuretic therapy. Plasma Na may be greater than 155 mmol^{-1}. Hyperthermia, seizures and focal central nervous system signs may occur in addition to impaired conscious level. The mortality in a recent series was 44% and was related to age above 60, uraemia and hyperosmolality but not to the degree or rate of fall of hyperglycaemia.

MANAGEMENT

1   0.45% saline can be alternated with 0.9% saline. Although plasma Na is high loss of water and Na is very marked and it is essential to replace some of the Na deficit. These patients have a greater fluid deficit than ketotic patients because of the longer history but as this occurs in older patients with an increased incidence of heart disease and renal impairment, rapid fluid replacement is fraught with danger. Pulmonary oedema and disequilibration may occur.

A CVP line is essential in management although it does not reflect preload on the left heart. Needs should be tailored to the specific patient and hypokalaemia corrected fairly urgently.

The alternating regime is probably safe unless plasma sodium is greater than 155 mmol l^{-1}, since plasma glucose should be falling. Plasma osmolality due to hyperglycaemia will also be falling and therefore some increase in plasma Na may act as a buffer to prevent too rapid a fall in osmolality which would precipitate cerebral oedema due to increased intracerebral water. Some workers prefer isotonic saline with its increased risk of Na overload relying on careful fluid balance, CVP and slower correction, because of the danger of hypotonic saline producing fatal cerebral oedema. The total Na deficit is likely to be about 400 mmol and water deficits are very variable (5–20 litres). A better guide to fluid replacement may be effective plasma osmolarity; 2(Na + K) + glucose + urea.

2   Less insulin will be required. 24 hour insulin requirements

rarely exceed 50 units. In this situation as glucose is taken up into the cell accompanied by water, intracellular volume increases at the expense of ECF. Following rehydration endogenous production of insulin may be adequate.

3 As has already been seen idiogenic osmoles are formed in the brain in situations of acute dehydration and this may be facilitated by insulin. Then isotonic water intoxication may be precipitated during fluid replacement therapy.

Potassium loss may have been extreme (5–10 mmol kg^{-1} body weight) and requires slow replacement. The risk of a precipitous, dangerous fall in K is greater in the absence of acidosis. It may be unwise to give insulin until plasma K is known to be greater than 4.0 mmol l^{-1}. K should be added to the infusion from the onset of treatment. 25–50 mmol may be required hourly. Oral K should be continued after the acute episode for one week because the total body K deficit may be as much as 400–1000 mmol.

If diabetic coma of any sort is present a nasogastric tube should be inserted to keep the stomach as empty as possible. A urinary catheter is useful but may lead to infection.

Subcutaneous heparin should be used as prophylaxis against deep vein thrombosis if the blood glucose levels are particularly high.

ECG monitoring is valuable to assess K status. Diagnosis and treatment of any underlying cause is extremely important. A search should be made for infection and the relevant specimens taken before starting the appropriate antibiotic.

The WBC count is raised in ketoacidosis without infection.

Lactic acidosis

This may be precipitated by phenformin. There is inhibition of oxidative phosphorylation and an increase in anaerobic glycolysis. The situation is worsened by renal insufficiency. There is a high mortality and large amounts of bicarbonate may be required for treatment. If ketosis is not marked but pH is low the anion gap will be raised. This subject is further discussed in Chapter 10.

Other circumstances in which blood sugar is raised include severe stress, burns, after subarachnoid haemorrhage, septicaemia and total parenteral nutrition.

DIABETICS REQUIRING SURGERY

This is one of the most important practical problems for housemen today. The precise management will be dependent on the extent of the surgery and whether it is elective or emergency but guidelines may be given.

Biguanides predispose to lactic acidosis and should be *stopped*.

Sulphonylureas. Long acting ones such as chlorpropamide should be stopped at least 3 days before operation and a change made to a short acting one such as glibenclamide for minor surgery or to soluble insulin for anything more extensive.

Preoperative preparation is aimed at avoidance of ketosis and hypoglycaemia. The latter may be unrecognized clinically under anaesthesia. Unstable diabetics may require 1–2 days in hospital to improve control. An assessment of renal function should be made as it is often impaired. Ketosis should be corrected and prolonged fasting avoided, hence the importance of operation early in the day on diabetic patients.

Insulin dependent diabetics

A continuous infusion of glucose and insulin should be set up 30 minutes preoperatively to give 2 units of insulin, 10 g glucose and 2 mmol K per hour. This may be continued until the patient is eating. Blood glucose and K levels needs monitoring and the concentration of either glucose or insulin modified as required. Many factors influence requirements, for example, the stress response to trauma, the extent of preoperative starvation and adequacy of control of diabetes preoperatively. An alternative is to give half the normal dose of insulin and set up an infusion of 5% dextrose at 200 ml hour^{-1}. The remainder of the insulin is given on return to the ward and glucose continued to give 3 litres 5% dextrose per day (150 g).

New developments

Administration of Hartmann's solution to diabetes intra-operatively is associated with a significant rise in plasma glucose compared to non-diabetics or alternative fluid administration. This is probably due to the lactate content (29 mmol l^{-1}). Lactate is an important gluconeogenic precursor especially in starved catabolic patients and may in part account for the increase in blood sugar. Saline is not contraindicated. Blood transfusion increases insulin requirement because although the glucose content of blood is low, lactate and pyruvate are high and these stimulate gluconeogenesis especially in diabetics. In the long term Mg and PO_4 replacement should be kept in mind.

In diabetic patients glucose induced hyperkalaemia is not uncommon and may be due in part to hypoaldosteronism and hyporeninism. Aldosterone is important in protecting against hyperkalaemia but not all diabetics with low aldosterone have glucose induced hyperkalaemia.

Human insulin is now available and is less immunogenic than other preparations. In the treatment of insulin dependent diabetics, a subcutaneous implanted i.v. insulin pump can improve blood glucose control in ambulant patients and the need for daily injections is eliminated. There is anxiety that these devices increase death rate due to malfunction of the device or bacterial endocarditis from the insertion site. There is probably no excess mortality but certain types of patient, such as those with autonomic neuropathy or renal disease may be at greater risk of death.

The Biostator is a machine which continuously samples venous blood, measures blood glucose and infuses insulin or glucose i.v. It has been used to aid control of brittle diabetics but requires an enormous amount of technical time.

An intranasal aerosolized preparation of insulin is now available with laureth 9 as surfactant. A dose of 1 mg kg^{-1} is absorbed in 15 minutes and reduces blood glucose by 50% in 45 minutes in fasting normal controls and by 50% in 120 minutes in fasting diabetics. Nasal irritation is very variable but it has been tried before meals as a supplement to ultralente insulin. In these circumstances it is well tolerated and glycaemic control is good as judged by glycosylated Hb. This may be a useful adjunct to subcutaneous therapy.

Hypoglycaemia

Hypoglycaemia is a danger of excessive or uncontrolled insulin therapy and should be treated with oral carbohydrate or in more severe cases by i.v. 50% dextrose. Many other conditions may be characterized by hypoglycaemia including anorexia nervosa. Glucagon plays a primary role in recovery from hypoglycaemia and adrenaline compensates for any deficiency in glucagon secretion. Both must be absent for failure to recover but patients on beta-blocking drugs have impaired recovery mechanisms.

POISONING

The most important aspect of treatment of drug overdose is early resuscitation and supportive care. Most drugs are metabolized in the liver and excreted in the urine. Drugs which are non protein bound are excreted by passive glomerular filtration. Rarely passive tubular diffusion also occurs.

Early excessive therapeutic zeal should be guarded against. For example gastric lavage is positively dangerous in corrosive poisoning, may cause water intoxication and increase absorption of the poison. Following initial resuscitation, measures to increase excretion of certain drugs may be appropriate. These measures include:

1 Forced diuresis.
2 Peritoneal dialysis.
3 Haemodialysis.
4 Haemoperfusion.

Forced diuresis

Most drugs are weak acids or bases which at the pH of the body exist partly as undissociated molecules. Drugs which are weak acids will ionize to a greater extent in alkaline solution. Cell membranes are more permeable to drugs which are lipid soluble and non-ionized. Therefore when a drug is ionized it fails to diffuse back into the circulation from the renal tubule and more will be excreted. Changing the pH of the urine to alter the degree of ionization therefore will effect the excretion of the drug. Only drugs which are excreted in the active form by the kidney will be affected. For acidic drugs the ionization will be increased in an alkaline urine and for basic drugs ionization is increased in an acid urine.

Forced diuresis will *only* be effective in the following circumstances:

1 When most of the drug is excreted in the toxic, unchanged form, in other words the drug is not metabolized.

2 The drug is distributed mainly in the ECF.

3 The drug is minimally protein bound.

4 The drug must be non-ionized at pH 7 but ionized at pH 8 (acidic drugs) or pH 6 (basic drugs).

The technique of forced diuresis involves transfusion of large amounts of fluid together with administration of a diuretic.

CONTRAINDICATIONS TO FORCED DIURESIS

1 Shock.

2 Impaired renal function.

3 Heart failure.

4 The elderly.

PROCEDURE FOR FORCED DIURESIS

Set up i.v. line taking blood for
    U & E
    sugar
    drug levels.

CVP line
Urinary catheter — measure urine pH
Arterial blood pH.

Forced alkaline diuresis

This is indicated for acidic drugs:

1 Phenobarbitone (plasma level > 100 mg l^{-1})

2 Barbitone (plasma level > 100 mg l^{-1})

3 Salicylates (plasma level > 500 mg l^{-1} or in a child 300 mg l^{-1}).

4 Lithium.

FORCED ALKALINE DIURESIS

1 Frusemide 20 mg i.v.

2 In first hour:
500 ml 5% dextrose
500 ml 1.4% $NaHCO_3$
500 ml 5% dextrose
3 If urine flow then less than 3 ml min^{-1} discontinue diuresis.
4 If urine flow more than 3 ml min^{-1} continue to maintain urine flow 500 ml $hour^{-1}$, giving boluses of frusemide 20 mg as required.
5 Add 10–20 mmol K to each litre of fluid.
6 Maintain urine pH 7.5–8.5 by adjusting amount of $NaHCO_3$ in infusion.

Recent work suggests that alkali alone is at least as effective as the potentially dangerous forced diuresis in enhancing salicylate removal and because of the need for caution, algorithms for modified alkaline diuresis have been devised.

Forced acid diuresis

This is indicated for weak bases which are normally partially ionized in solution.
1 Quinine.
2 Amphetamine.
3 Fenfluramine.
4 Monoamine oxidase inhibitors.
5 Phencyclidine.

FORCED ACID DIURESIS

1 Frusemide 20 mg.
2 In first hour
1000 ml 5% dextrose
+ 500 ml 0.9% NaCl
+10 g arginine (or lysine) hydrochloride.
(Intravenously over 30 minutes.)
3 If urine flow then less than 3 ml min^{-1}, discontinue diuresis.
4 If urine flow more than 3 ml min^{-1} continue to maintain urine flow 500 ml $hour^{-1}$, giving boluses of frusemide 20 mg i.v. as required.
5 Add 10–20 mmol K to each litre of fluid.
6 Maintain urine pH 5.5–6.5 with oral NH_4Cl 4 g 2 hourly.

During forced diuresis measure:

Urine pH hourly.
Plasma and urine electrolytes 4 hourly.
Keep accurate fluid balance chart.

COMPLICATIONS

Forced diuresis requires the patient to be in an intensive care unit
as very great care should be exercised with this technique. In spite
of this complications include:

1 Water intoxication.
2 Cerebral oedema.
3 Electrolyte and acid base disturbance.

Peritoneal dialysis

This technique has even less application in the treatment of drug
overdose although it has been used for poisoning with lithium and
ethylene glycol.

Haemodialysis

This technique is the treatment of choice for severe poisoning with
lithium, methyl and ethyl alcohol. Very rarely it may be useful in
removing phenobarbitone, barbitone and salicylates if the
patient's renal function is impaired so that forced alkaline diuresis
is contraindicated.

Haemoperfusion

This technique involves passing the patients blood over an
adsorbent such as charcoal or resin which has been coated with a
substance such as cellulose to prevent platelet and fibrinogen
consumption.

Indications for haemoperfusion

1 The drug should be readily taken up onto the adsorbent.
2 Much of the drug should be in equilibrium with plasma water.

3 The blood level of the drug should be directly related to its toxic effects.
4 The technique must significantly improve on the body's normal mechanisms of elimination of the drug.

If the drug is non-toxic or has a large volume of distribution, is irreversible in its action or has an antidote then haemoperfusion is contraindicated.

Heparinization will be required which may lead to haemorrhage. In addition loss of platelets, white blood cells and clotting factors are recognized complications. The plasma concentrations of calcium, glucose, urea, creatinine and urate fall.

Details of abnormalities to be found in specific drug overdose are beyond the scope of the book but relevant characteristic features do occur in salicylate overdose.

Salicylate overdose

1 Stimulation of the respiratory centre causes hyperventilation with a fall in $PaCO_2$ and an increase in pH. Renal compensation reduces reabsorption of HCO_3 producing a fall in plasma HCO_3.
2 Later metabolic acidosis develops with ketonaemia.

Cyanide poisoning

Acute cyanide poisoning causes histotoxic hypoxia and profound lactic acidosis. Sodium nitrite, cobalt edetate, hydroxycobalamin and 100% oxygen are appropriate forms of treatment together with sodium thiosulphate.

CENTRAL NERVOUS SYSTEM DISEASE

In the brain the cerebrospinal fluid (CSF) is in very close association with ECF. If K, Mg or Ca is infused intravenously into animals there is little change in cisternal CSF concentration of the ions infused. Therefore entry of these ions into cerebral ECF is slow at least and powerful mechanisms in the blood brain barrier (BBB) and choroid plexus exist which keep the cerebral ionic content constant. Water is, however, freely permeable between brain and systemic ECF and therefore changes in plasma osmolality are rapidly transmitted to the central nervous system. In hypernatraemia where water is lost from the cells the brain is able to create idiogenic osmoles which 'hold' water within the brain.

The BBB and blood CSF barriers are freely permeable to CO_2 but much less permeable to HCO_3 and cerebral ECF has a relatively low buffering capacity. Brain pH is very stable. Small changes in brainstem CO_2 alter ventilation to restore pH to normal. Although the BBB actively transports HCO_3, changes in the concentration of this ion occur slowly within the brain.

Abnormal osmolality and the brain

Hypernatraemia associated with hyperosmolality is usually due to water deficiency and caused by an impaired conscious level with reduced voluntary intake in response to thirst. Hypernatraemia may cause physical damage to the brain. Since the brain is situated within the closed cranium sudden expansion or contraction of one compartment will effect the other compartments. Administration of mannitol may produce such fluid shifts that in infants a subdural haematoma due to tearing of the delicate veins occurs as the cerebral substance shrinks.

If dehydration is corrected too rapidly further osmotic swelling may occur within the brain cells. In this context, administration of 5% dextrose is akin to giving water. Water intoxication producing osmotic swelling of the brain is distinct from cerebral oedema due to trauma, neoplasm or infection when there is an increase in Na and loss of K as well as increased water. Hyponatraemia does not seem to be associated with such a risk of permanent brain damage as hypernatraemia. Rapid reduction in blood urea during dialysis may produce cerebral symptoms of water intoxication because exchange of urea across the BBB is slow and therefore the brain will be hypertonic in relation to the rest of the body and as such will take up water.

Causes of hyponatraemia of cerebral origin

1 Meningitis especially tuberculous.
2 Severe head injury.
3 Cerebral tumour.
4 Subarachnoid haemorrhage.
5 Hypertensive encephalopathy.

6 Encephalitis.
7 Poliomyelitis.
8 Cerebral abscess.
9 Guillain Barré syndrome.
10 Acute intermittent porphyria.

These situations are due mainly to high circulating ADH hormone levels (see Chapter 6). Treatment is that of the underlying cause and water restriction.

Management of cerebral oedema

Management of cerebral oedema hinges on the following:

1 Maintenance of good oxygenation.
2 Hyperventilation to reduce cerebral blood flow.
3 CSF drainage. It is mandatory to remember that ill advised lumbar puncture in the presence of raised intracranial pressure may precipitate coning with fatal consequences. CSF drainage should therefore be effected from the ventricular system in these circumstances.
4 Hypertonic solutions. Mannitol exerts a profound osmotic diuretic effect and reduces brain volume.
5 Frusemide in addition to its effect as a loop diuretic also reduces the rate of CSF production.

DISORDERS DUE TO HEAT

Heat exhaustion, heat stroke

This occurs when insufficient water is taken to replace the losses due to sweating. Normally, excessive thirst will lead to an increased voluntary intake. However, if water is not available or the patient has an impaired conscious level then heat exhaustion is likely. It is commonest in the elderly, those with cardiovascular disease or alcoholics. Marked hypernatraemia and hyperosmolality occur. Heat stroke results in metabolic acidosis often accompanied by respiratory acidosis. Hypocalcaemia and less commonly hypophosphataemia may occur. Treatment consists of

5% dextrose intravenously or if the diagnosis is not clear and salt loss may have occurred then 0.9% saline is suitable.

Heat exhaustion due to salt depletion occurs when salt is inadequately replaced during long term sweating. Fatigue and anorexia occur with vomiting and circulatory collapse. The salt content of sweat is about 45 mmol l^{-1} in these circumstances and the deficit may be calculated (see Chapter 7). This situation develops over a period of weeks. Treatment consists of giving salt to unacclimatized patients.

Fatal heat stroke occurs in long distance runners. Mortality (30%) is directly related to temperature. At high temperatures sweat glands do not respond to acetylcholine and heat loss is inadequate despite increased skin blood flow. Liver damage leads to impaired haemostasis, occasionally overt disseminated intravascular coagulation and fibrinolysis. Acute renal failure occurs in 25% of cases. Adequate rehydration is essential. Anhidrotic heat exhaustion occurs after several months residence in a hot climate. Defective sweating occurs with an increase in body temperature. Potassium depletion may be important in the aetiology in which case supplements should be given. When heat stroke occurs the best treatment is rapid cooling by evaporative heat loss from warm skin. Hence, it is important to maintain peripheral vasodilatation and then treat shock, fluid depletion and electrolyte disturbance and especially acidosis.

Malignant hyperthermia

This syndrome consists of a rapid rise in temperature (in excess of 2 °C hour^{-1}) with rigidity and cyanosis precipitated by exposure to some anaesthetic agents in particular the combination of suxamethonium and halothane. Early changes which occur are a fall in pH, a rise in $PaCO_2$ and K.

BURNS

The essential problem in burns is the rapid fluid loss from the circulation. Up to 50% of the circulating blood volume may be lost in the first 6 hours and as the loss is diffuse and difficult to recognize or assess it may be underestimated with subsequent dangerous delay in treatment.

The local lesion

The oedema fluid has much the same electrolyte content as plasma with a protein content of 40–50 g l^{-1}. The protein in interstitial burn fluid comes from circulating plasma because of the increased permeability of the damaged capillaries. This reduces plasma protein osmotic pressure and increases interstitial fluid in the non-burned parts of the body. Much of the Na lost into interstitial fluid enters cells whereas K is lost into ECF and plasma. The increase in capillary permeability is maximal during the first 8 hours after the burn and gradually decreases over the next 36 hours. Hypoxia in the damaged area may increase anaerobic metabolism with metabolic acidosis and this will further impair tissue perfusion.

Widespread damage to blood vessels causes thrombosis and red blood cell aggregation. For each 10% of the body surface burned there is a 7% reduction in red blood cells.

Because of the loss of intravascular protein there is a reduction in effective plasma volume. Further loss of fluid occurs due to the longer lasting persistent evaporation of water from the damaged skin surface. This loss varies with the temperature and humidity of the environment but can be 3 litres daily. Large heat losses will occur due to latent heat of vaporization.

Systemic disturbances

Hypovolaemia may be severe enough to produce shock. There will be oliguria and a reduction in cardiac output which may be associated with the production of myocardial depressant factor. Catecholamine release increases peripheral resistance. The loss of plasma is in excess of the loss in red cell mass so that haematocrit rises. There is an increase in viscosity with sludging of red blood cells in the microcirculation. This produces further hypoxia and metabolic acidosis. Red blood cells may release large quantities of pigment which when deposited in the renal tubules will further impair renal function. Polyuric renal failure is not uncommon with low fractional excretion of Na, reduced GFR and a proximal tubular defect. The distal nephron often remains intact and functioning.

The metabolic response to trauma is activated. Hypercatabolism leads to increased K loss and subsequently if untreated to hypo-kalaemia. Rarely can sufficient calories be given either orally or

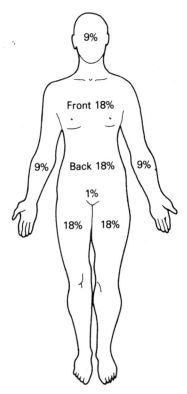

Fig. 14.2 Rule of nine for calculation of the percentage body surface area burn in an adult.

parenterally to prevent breakdown of the patient's own muscle protein.

Inhalational damage may occur. Heat damage rarely extends beyond the trachea but inhalation of chemicals such as aldehydes may produce pneumonitis. Adult respiratory distress syndrome due to sepsis or oxygen toxicity may occur when pulmonary capillary permeability is markedly increased.

Gastrointestinal ulceration may occur as may paralytic ileus with all its attendant electrolyte problems.

Management

Many formulae exist for fluid replacement in burns. These should be used as a guide only. The first priority is early fluid replacement.

The intravenous infusion of fluid should start as soon as possible to prevent the fall in cardiac output and to maintain renal perfusion. If a patient with burns develops renal failure his chances of survival are considerably reduced.

Most formulae depend on replacement according to the percentage of skin burned.

Burns in small children do not conform to this rule exactly since the head and neck represents a greater proportion of the surface area in this age group.

One commonly used formula for all burns is as follows:

$$\frac{Body\ weight\ (kg)}{2} \times \%\ area\ burned = \text{ml of fluid required in first}$$

$$\text{four hours}$$

The work of Muir and Barclay divided the first 36 hours after a burn into period of:

1   4 hours.
2   4 hours.
3   4 hours.
4   6 hours.
5   6 hours.
6   12 hours.

The expected fluid requirement in each of these periods is given by the formula. Most of the fluid requirements may be given as plasma but as we have already seen red cells are lost so that blood should be given as part of the fluid replacement. One formula for giving blood is to give 1% of the patients normal blood volume (70–75 ml kg^{-1}) for each percentage full thickness burn.

In an adult a burn of greater than 15% requires intravenous infusion but in a child this will be required with a burn of 10% surface area. Burns in excess of 50% may need more fluid and formulae should be used as a guideline only especially in the very young and very old.

The clinical state of the patient is a good guide to management but additional monitoring is very important and the following are essential:

1   Urine output.
2   Urine osmolality, plasma osmolality.
3   Haemoglobin, haematocrit (colloid resuscitation only).
4   CVP.
5   Peripheral core temperature difference.

Urine output depends not only on the renal perfusion and state of the circulation but also on the water retention that occurs when the metabolic response to trauma is activated. What is important is that the volume of urine excreted should be sufficient at whatever osmolar content to excrete the body's osmolar load. In conditions of stress and trauma with maximal ADH activity the concentration of urine is less than 1000 mosmol kg^{-1}. An average normal osmolar output of 1200 mosmoles therefore would require a urine volume of at least 1200 ml for its excretion or 50 ml hourly. If the urine output is low but osmolality high this indicates poor renal perfusion and is an indication to increase the rate of transfusion. A good urine volume alone, however, is of little value if its osmolality is low. This implies impaired renal function. In a severe burn hourly urine volume must be measured accurately and a urinary catheter is justified.

Maintenance of circulating blood volume and urine output therefore are the aims of fluid therapy. The use of blood and plasma have already been referred to. Other plasma expanders may also be used in limited volumes (see Chapter 11). The use of crystalloid solutions such as Hartmann's will often produce a good urine output but its effect on resuscitation as judged by the CVP is minimal. Measurement of the haematocrit is a useful indication of the plasma deficit in many circumstances but in burns it must be remembered that there is a variable loss of red blood cells and the significance of the haematocrit after transfusion of whole blood is difficult to determine.

Low volume colloid resuscitation regimens are associated with very high levels of plasma renin activity and aldosterone especially in the first 5 days post-burn with marked Na retention. Secondary peaks in the levels of these two hormones is associated with deterioration and systemic sepsis. Attempts to fill plasma volume with isotonic crystalloid solutions only succeed at the expense of producing interstitial oedema in unburned parts of the body. Colloid administration on the other hand, results in minimal non-burn oedema with more rapid return of the cardiac output to normal. About 0.5 mmol kg^{-1} Na are required for each percentage burned body surface area. Evaporative loss from burns can be very high (1–21 per 9% burn in the first 24 hours). Failure to replace this results in progressive plasma hyperosmolality although too much Na free water in the presence of increased ADH will give rise to hyponatraemia and cellular overhydration. The aim therefore is modest ECF hypernatraemia produced by colloid or possibly hypertonic saline.

The burned patient will have ongoing fluid losses and later considerable blood loss may occur during repeated surgery for desloughing. In an inhalation burn in which ARDS develops, plasma transfused intravenously may appear in pulmonary aspirate. This situation is very difficult to treat. In addition to specific fluid replacement other factors deserve attention. The humidity of the environment should be increased along with the temperature in order to reduce water and heat loss and hence energy expenditure. Treatment of pain with generous doses of narcotic analgesics will reduce the stress response. Parenteral nutrition may be required to provide adequate energy and protein needs.

Burned patients often receive repeated large blood transfusions and are one of the main groups at risk for non-A, non-B hepatitis.

EFFECTS OF DRUGS
ON ELECTROLYTE BALANCE

Drugs can interfere with normal uptake, elimination, regulation and body distribution of electrolytes. These effects are likely to be greater in the elderly as renal function begins to deteriorate at 40 with reduction in GFR and renal tubular secretion of about 1% annually. Plasma creatinine is not a very good indication of renal function because muscle mass and therefore creatinine production is reduced in this age group. Drugs disposed of by renal excretion will accumulate. Older people are more sensitive to benzodiazepines, anticoagulants and diuretics than the young.

Diuretics

Acetazolamide inhibits carbonic anhydrase activity in the proximal tubule thereby blocking exchange of H for Na there. Plasma H increases and pH falls and therefore this may be valuable in the treatment of metabolic alkalosis. Plasma HCO_3 and Na fall and hyperchloraemia occurs because Cl is reabsorbed in the ascending limb of the loop of Henle. Na is exchanged for K in the distal tubule resulting in hypokalaemia.

Thiazide diuretics selectively inhibit reabsorption of Na and Cl in the early distal tubule so that plasma Na and Cl fall. There is an increase in exchange of Na with K and H in the distal tubule so that hypokalaemia (which may play a part in genesis of dysrhythmias and sudden death) and metabolic alkalosis may occur with an increase in plasma HCO_3. Thiazides produce a steady increase in plasma urate during prolonged use in hypertension. Renin and aldosterone are increased and plasma Na and K reduced. Glucose intolerance develops, Ca and cholesterol increase and high density lipoproteins fall. In moderate or severe hypertension, however, the benefits outweigh the risks.

Frusemide and ethacrynic acid inhibit Na and Cl reabsorption in the ascending limb of the loop of Henle. Both are potent diuretics and produce a fall in plasma Na, Cl and volume. There is exchange of Na with K and H in the distal tubule with a fall in plasma K and metabolic alkalosis. When loop diuretics are used in moderate doses for control of mild CCF they are less likely than thiazides to cause hypokalaemia. Differential diagnosis from dilutional hyponatraemia may be difficult but blood volume, CVP, plasma proteins and haematocrit may help to differentiate (see Chapter 7). Side effects are similar to those of the thiazides, both causing hyponatraemia. Rapid diuresis can precipitate urinary retention or incontinence. Frusemide causes acute venodilatation with reduction in left ventricular filling pressure within minutes of administration and before the diuretic effect is apparent due to an effect on angiotensin II.

Spironolactone is an aldosterone antagonist acting on the distal tubule so that sodium is not reabsorbed in exchange for H and K and a small increase in plasma K and H occur with a small fall in plasma sodium.

Triamterene and amiloride act similarly although they are not aldosterone antagonists.

Mannitol is an osmotic diuretic producing hyponatraemia and hypokalaemia with repeated use. Its main use is in treatment of raised intracranial pressure and in some cases of impending renal failure.

Hyponatraemia occurs with chronic diuretic therapy. Salt depletion can reduce renal blood flow and precipitate lithium toxicity. Hypercalcaemia and hypocalciuria occur during long term treatment with thiazide diuretics. Loop diuretics, K sparing diuretics and spironolactone increase Ca excretion.

Glucocorticoids

In primary and secondary adrenal insufficiency with lack of hydrocortisone, even with a normal mineralocorticoid activity there is inability to excrete a water load and susceptibility to water intoxication. In primary hyperaldosteronism and Cushing's syndrome with increased cortisol production severe hypokalaemia may occur due to increased K loss in the urine. Raised cortisol levels are known to be associated with hypokalaemic alkalosis. Patients with Cushing's syndrome may have an expanded plasma volume and occasionally mild oedema. Usually plasma sodium is normal.

Many drugs may influence water and electrolyte composition of the body. For example carbenoxolone and liquorice derivatives gives rise to sodium retention. Large doses of parenteral antibiotics may give a large sodium load. Certain electrolyte solutions for parenteral administration are incompatible with certain drugs; for example tetracyclines will precipitate in Hartmann's solution (because of its calcium content). Erthyromycin should not be added to 5% dextrose solution. The activity of lignocaine and mexitiline (type I antiarrythmic drugs) is reduced in the presence of low ECF K.

The nature of parenteral fluids themselves is also important. Large volumes of 5% dextrose (pH 4.0) used as a vehicle for infusion of drugs for example may produce metabolic acidosis and large volumes of 8.4% sodium bicarbonate (1 mmol ml^{-1}) will provide a large salt load which is especially dangerous in congestive cardiac failure.

PREGNANCY

Maternal plasma volume increases progressively during the second and third trimester of pregnancy and is greatest in multiple pregnancy. Maternal plasma volume falls in pre-eclamptic toxaemia (PET) and chronic hypertension. In PET low plasma albumin is probably due to urinary loss and this predisposes to further oedema. PCV reflects plasma volume provided there is no bleeding or anaemia. Viscosity is increased in PET and plasma expanders may improve organ perfusion although at some risk of generalized increase in oedema.

At term ECF volume is almost always expanded. Glucose and

oxytocic infusions should be used with care or hyponatraemia and water intoxication may occur. Fluids are required as a vehicle for oxytocin, to prevent maternal ketosis and during maintenance of epidural analgesia, to prevent hypotension.

CANCER

There are several indirect metabolic manifestations of cancer which may affect fluid and electrolyte balance.

Cushing's syndrome.

Syndrome of inappropriate ADH production.

Hypercalcaemia.

Cachexia.

Hypoglycaemia.

Zollinger Ellison syndrome.

Glucagonoma.

Hormone secretion may be related to the effects of the tumour or synthesis of ectopic hormones. The major problem, however, is the anorexia of cancer. Further investigation of methods of nutritional support are required.

Chapter 15
Practicalities of Setting Up
Infusion Lines

About 20 million intravenous infusions (i.v.i.) are set up annually in the United Kingdom.

PERIPHERAL INFUSIONS

Insertion of a peripheral intravenous line is now commonplace in medical practice. It may be a life saving procedure in cases of hypovolaemic shock or more simply a means to supply adequate fluid therapy for the patient. In any event skilful insertion of a cannula with the minimum of complications should be the aim of all junior doctors.

Equipment

1 The infusion fluid should be checked by the doctor for the following points:
(a) Is the fluid clear?
(b) Is the fluid correct in type and concentration and does it correspond with the prescription chart?
2 The infusion fluid and administration set should be assembled aseptically and all air bubbles excluded from the system. The risk of air embolus is greatly increased in the presence of a congenital right to left intracardiac shunt.
3 Skin preparation. This does not totally prevent bacterial contamination. One of the alcohol preparations is suitable for example chlorhexidine in 70% alcohol.
4 Cannulae.
5 Razor.
6 Clean paper towels and waterproof sheet.
7 Fixative.

Choice of cannula

In an adult the smallest cannula likely to be adequate is an 18
gauge. For large infusions one or more 14 gauge cannulae may be
used. To some extent choice of cannula is a personal preference.
However, polyvinylchloride (PVC) cannulae are more irritant
than Teflon ones. Some controversy exists regarding the use of a
cannula with a side port for injection. This injection port may act
as a sump which cannot be cleaned prior to the injection procedure
and a 55% incidence of contamination with pathogenic organisms
at this site has been reported.

The origin of organisms inside the injection port does not
matter, but what does matter is that these may be injected and
produce systemic sepsis. Catheter related thrombosis occurs in up
to 45% of cannulations and there is a significant correlation
between this and positive catheter tip cultures. The incidence of
thrombosis is reduced by low dose heparin and bacterial filters,
scrupulous attention to technique including use of 0.5% chlorhexidine in spirit and increased awareness of the problem. In health a
few bacteria in the bloodstream may be harmless but in an ill
patient, especially one with a damaged heart valve or immunosuppression, this is clearly unacceptable. Despite this, one such cannula with a side port (Venflon) has shown a similar incidence of
phlebitis (35%), positive cannula tip culture (15%) and bacteraemia (1%) as other cannulae used for continuous flow infusion.
Nevertheless there is pressure on companies to discontinue manufacture of devices with an injection port which cannot be safely or
repeatedly disinfected.

Blood transfusion increases the risk of subsequent infection by
formation of a fibrin sheath on cannulae. Thus peripheral cannulae
should be replaced after blood transfusion. Precautionary
cannulae should be avoided.

Procedure

The doctor should explain the procedure to the patient and
reassure him. The most suitable veins are those of the forearm
avoiding sites close to the wrist and elbow since flexion may
obstruct the flow of the infusion. It is advisable to remove the arm
from the gown or other clothing and to shave the area if the patient

is especially hirsute. A tourniquet or sphygmomanometer cuff can be used to obstruct the venous flow. The veins should be palpated. The junction point between two veins is a convenient access point.

The hands are washed and dried on a clean towel. Then the skin is prepared. 1% lignocaine may be injected through a fine needle to raise a wheal and provide local anaesthesia. Unfortunately this may to some extent obscure the anatomy. The vein should be stabilized with the left hand by traction on the skin towards the patients fingers. The cannula and needle are then inserted through the skin a few millimetres distal to the vein. As the vein is entered blood flashes back into the cannula. At this point the cannula can be advanced into the vein as the needle is withdrawn. The tourniquet is deflated and the infusion system connected. Blood spillage should be reduced to a minimum by pressure on the cannula tip within the vein as the needle is withdrawn and the infusion connected.

Fixation

Adhesive tape should be used to ensure fixation. The infusion tubing may be looped to the skin then the whole area with the exception of the injection port may be covered with broad elastoplast avoiding total encirclement of the limb.

Problems

1   *Venous access*. It is not uncommon in an ill patient to find many damaged thrombosed veins due to previous infusions. In these circumstances it is helpful to ensure a vasodilated limb if necessary by placing the arm in warm water. A sphygmomanometer cuff accurately pumped up to between systolic and diastolic pressure will help. The patient is encouraged to clench his fist several times and gentle tapping over the vein will cause it to stand out. As far as possible veins on the dorsum of the hand should not be used. If an antecubital fossa vein is the only suitable one then insertion of a long central venous line should be considered. The basilic vein is a useful, often overlooked, site.

2   *Thrombophlebitis*. The puncture site becomes contaminated with bacteria and then infection within the lumen of the cannula extends to the vein. This is made worse if irritant fluids or drugs

are infused or if infusion is prolonged. In most circumstances the bacteria arise from the patient's own skin and not from the infusion fluid. The use of bacterial filters to remove organisms from the infusion fluid has little effect on the incidence of thrombophlebitis. The cannula should be removed and a new one sited elsewhere. Injection ports, cannula hubs and 3-way taps are readily contaminated with gram negative organisms.

Any procedure which involves handling the system, for example addition of drugs to the i.v. fluid, changing the giving set or flushing an occluded line is associated with an increased risk of contamination. Filters trap fungi and most bacteria. Contamination of the actual fluids can rarely occur during manufacture or use.

3 *Infusion fluid.* Having sited the cannula it is essential to ensure that the fluid regimen is clearly written up on the prescription chart and that it is appropriate for the patient. One of the commonest problems in management of i.v.i. is dehydration or overhydration. Volume overload is easy to achieve with injudicious fluid prescription especially in the elderly with impaired cardiac and renal function.

4 *Deep vein thrombosis.* It has been shown by Janvrin that blood clots more readily when diluted with saline. Recently a trial in 60 patients undergoing routine laparotomy showed that 30% of the patients given an i.v.i. sustained a deep vein thrombosis (DVT) as assessed by radioisotope scanning using ^{125}I fibrinogen uptake whereas only 7% of the patients who were not transfused had a DVT.

For most purposes of fluid administration manual drip regulation is considered adequate. However, in paediatric practice, or in adults where fluid restriction is appropriate and for more accurate methods of fluid and drug administration automatic methods are required. One solution to this problem is to use a burette system loaded with the correct fluid volume to be infused over the next hour. An alternative is to use an infusion pump. These fall into two categories:

(a) Drip counters
(b) Volumetric infusion pumps.

Only volumetric infusion pumps are really accurate at low volumes and overcome the persistent problems of changes in venous pressure, drip rate and viscosity. Also available are motorized syringes designed to give variable volumes over the number of hours. These are useful for continuous heparin or insulin administration but because of their size (50 ml usually), they cannot cope with large volume infusions. Drip counters are dependent on drop size which will vary with the administered solution.

The importance of thorough mixing of drugs added to an infusion cannot be overemphasized. KCl and heparin both tend to form a pool if inadequately mixed in PVC containers. Uneven distribution is potentially lethal for the patient. Insulin floats to the top of haemaccel containers with similar disastrous consequences. If insulin is added to the plastic i.v. infusion bag the needle through which it is inserted must be long enough to traverse the injection port dead space otherwise the insulin will never reach the infusion fluid. Addition of drugs to i.v. fluids and their subsequent administration is a difficult task and requires reappraisal and improvement.

CENTRAL VENOUS PRESSURE CANNULATION

There are a few emergency circumstances where absence of peripheral veins necessitates cannulation of a central vein as a life saving procedure. More often, however, this is an elective undertaking.

Indications

Monitoring

Central venous pressure (CVP) varies with:
1 Venous blood volume.
2 Right ventricular function.
3 Venous tone.
Existing or anticipated large changes in blood volume are the commonest reason for CVP monitoring.

Administration of substances irritant to peripheral veins

Solutions of amino acids and concentrated carbohydrate used for parenteral nutrition are hypertonic, irritant to peripheral veins and must therefore be given into a central vein. Inotropic agents such as adrenaline produce profound constriction if administered through a peripheral vein. 25–50% Althesin infusions produce severe peripheral thrombophlebitis and should therefore be given into a central line. However, at least one case of subclavian vein thrombosis occurred.

Equipment

Insertion of a CVP line should be done with strict asepsis. A special pack is a great help and should contain ample towels and swabs.
Skin preparation. An iodine preparation in spirit is suitable.

Catheters

Many varieties are available. The only recommended ones are those which use a cannula over a needle which is inserted into the vein, the needle withdrawn and the catheter threaded through the cannula. If a catheter is threaded directly through a sharp needle only a tiny amount of catheter withdrawal may result in shearing off the catheter. PVC catheters are relatively hard and have been associated with perforation of the right atrium or ventricle with fatal cardiac tamponade. Teflon catheters are less irritant and silicone coated ones are least likely to be associated with fibrin formation and subsequent bacterial colonization. These are particularly suitable for long term parenteral nutrition. Use of a Seldinger technique for insertion is also recommended.

Choice of site

The antecubital fossa veins may be used. The median basilic vein is preferable as a long catheter threaded via the cephalic vein will sometimes not go beyond the level of the clavipectoral fascia. The external jugular vein is not a recommended site as only 50% of catheters can be threaded to an intrathoracic position. The internal jugular vein has the advantage of being straight but its close proximity to the carotid artery is a disadvantage. The infra-clavicular approach to the subclavian vein is popular for long term cannulation as fixation is easy and the patient can be very mobile.

No attempt should be made to insert a cannula at a septic site. In the septicaemic patient a central catheter may provide a continuous source of bacteraemia. Blind cannulation of an invisible vein in the neck is very dangerous and contraindicated in disorders of coagulation. There is a danger of air embolus when the catheter enters the great veins in a hypovolaemic patient. Measures to avoid this are essential.

Arm veins

Essentially the method is initially that for setting up a drip. The catheter should be 70 cm or more in length for an adult and as it is threaded the head should be turned to the same side to compress the neck veins. When the tourniquet is released the proximal end of the catheter should be kept low to avoid air embolism. About 75% of catheters should be accurately placed using this technique.

Internal jugular vein

This vein runs behind the sternomastoid close to the lateral aspect of the carotid artery.

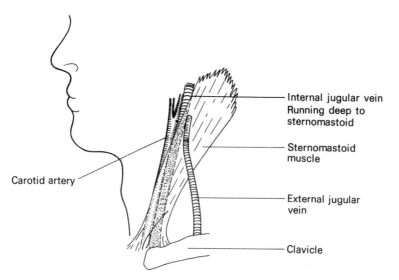

Fig. 15.1 The position of the internal jugular vein.

The right internal jugular vein is easier for the right handed operator and also avoids injury to the thoracic duct. The patient is placed 20° head down with his head turned away from the side of cannulation. The point of entry is the midpoint of a line joining the mastoid process to the sternoclavicular junction. The carotid artery should be gently pulled medially and the needle with its cannula is attached to a syringe of saline. The needle should be at 30° to the skin and directed towards the feet slightly laterally from the midline. As the needle is advanced aspiration will confirm entry into the vein. The cannula is then inserted, the needle withdrawn and the catheter threaded through the cannula. It can then be connected to a bubble free infusion system. This route is associated with the highest number of correct placements (up to 99%).

Subclavian cannulation

The infraclavicular approach to the subclavian is now a popular site for CVP insertion.

The subclavian vein crosses the first rib behind the medial third of the clavicle to come intrathoracic. The patient is positioned as

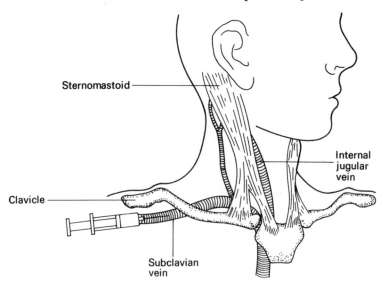

Fig. 15.2 The position of the subclavian vein.

for internal jugular cannulation. A point 1–2 cm below the mid-point of the clavicle is marked.

Using a long needle and cannula attached to a saline filled syringe the assembly is advanced along the under surface of the clavicle in the direction of the sternoclavicular junction. Aspiration of dark venous blood confirms successful subclavian puncture. The procedure is continued as already described. It may be necessary to depress the head of the humerus in order not to go too deeply. Occasionally a sandbag between the shoulders is helpful. This route may be associated with a rather high rate of catheter misplacement (5–25%). However, fixation is much easier and allows greater patient mobility during long term use. Subclavian catheterization of small children (<10 kg) is greatly facilitated by use of the Seldinger technique which uses a guidewire.

External jugular

This vein can be used under some circumstances but requires a J wire for placement in order to negotiate the external jugular subclavian junction.

Procedure

The patient should be reassured and a full explanation of the procedure given. The operator should scrub up and wear gown, mask and gloves. The skin is then cleaned over a wide area and the chosen site surrounded with sterile towels. Local infiltration with 1% lignocaine is used unless the patient is receiving general anaesthesia.

Confirmation of catheter position

When infusion fluid has been connected it should be lowered towards the floor. Back flow of blood confirms the intravascular position of the catheter tip. When the catheter is connected to a fluid manometer for recording the value of the CVP, oscillations of the fluid column in time with respiration and sometimes pulse rate confirm the intrathoracic position.

However a chest X-ray is mandatory following even an unsuccessful attempt at cannulation for two reasons:

(a)   To confirm the position of the catheter tip.

(b)   To exclude a pneumothorax. for this reason the chest X-ray should be taken in expiration.

Fixation

Many systems include an attachment for skin suture which is invaluable for long term use in preventing accidental removal. An Opsite or similar plastic dressing which is transparent will allow inspection of the wound without additional manipulation, which may lead to contamination. The catheter must be very firmly attached to the skin.

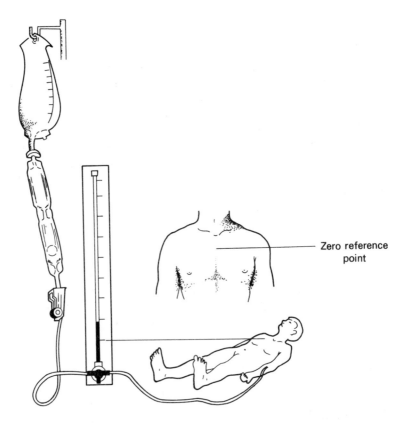

Zero reference point

Fig. 15.3   Central venous pressure manometer.

Following insertion strict asepsis is necessary to avoid contamination. Drugs should preferably be injected through a peripheral line and taps excluded from the central line since these are a site of potential introduction of bacteria.

Zeroing

The author uses the sternal Angle of Louis as a reference point. Others prefer the anterior or midclavicular line. It is essential that the reference point is clearly defined. The normal CVP as measured from the Angle of Louis is −3 to +3 cm water.

Problems

Immediate

1 Damage to large vessels in the neck. Carotid or subclavian artery puncture may occur accidentally. Firm pressure should be applied to the carotid if this is punctured keeping a watch on the pulse rate for bradycardia due to carotid sinus pressure. Veins may also be perforated resulting in substantial haemorrhage.
2 Pneumothorax, haemothorax. Either of these may be fatal. The importance of the post-insertion chest X-ray cannot be over-emphasized.
3 Thoracic duct injury.
4 Phrenic nerve damage.
5 Failure to place the catheter correctly.
If misplaced catheters are not recognized irritant fluids may be infused extravenously resulting in tissue damage. A Doppler ultrasound probe may help to locate subclavian vessels and minimize trauma. The commonest misplacement is upwards into the ipsilateral internal jugular vein where infusion of hyperosmolar TPN solutions has resulted in spreading cerebral cortical vein thrombosis.

Throughout

1 Air embolus
2 Catheter detachment.
3 Cardiac dysrythmia.

Late complications

1 Sepsis including bacterial endocarditis.
2 Central vein thrombosis. This is commonest during prolonged cannulation for parenteral nutrition.
3 Myocardial perforation and tamponade.
The overall complication rate for neck vein cannulation is about 15%. Unexpected fever for 12 hours is an indication to remove the central line and culture its tip. Some organisms seem able to survive exposure to bactericidal concentrations of some antibiotics when adherent to PVC catheters. Scrupulous attention to detail is essential to reduce the complication rate in this increasingly commonly performed procedure. Stopcocks, taps and integral side ports are a potential nidus for pathogenic organisms as discussed for peripheral lines but the consequences of infection in central lines are much more serious.

Cannulation of peripheral arteries and insertion of a Swan Ganz catheter are beyond the scope of this book. Information is supplied elsewhere.

Further Reading

General

BERLYN G. M. (1980) *A Course in Clinical Disorders of Body Fluids and Electrolytes*. Blackwell Scientific Publications, Oxford.

CAROL H. J. and OH M. S. (1978) *Water, Electrolyte and Acid Base Balance; Diagnosis and Management*. Kippincott, Philadelphia.

HODSMAN G. P. and ROBERTSON J. I. S. (1983) Captopril: five years on. *Brit. Med. J.* **287**, 851–2.

LAWRENCE C. A. (1981) Ion-selective electrodes. *Brit. J. Hosp. Med.* **26** 625–30.

SKORECKI K. L. and BRENNER H. M. (1981) Body fluid homeostasis in man. A contemporary overview. *Am. J. Med.* **70**, 77–88.

VIDT D. G., BRAVO E. L. and FOUAD F. M. (1982) Medical intelligence, drug therapy, Captopril *New Engl. J. Med.* **306**, 214–9.

Specific

Water and Salt

BOON N. A. and ARONSON J. K. (1985) Dietary salt and hypertension; treatment and prevention. *Brit. Med. J.* **290**, 949–50.

EDITORIAL (1984) Diet and hypertension. *Lancet* **ii**, 671–3.

FLEAR C. T. G. and GILL G. V. (1981) Hyponatraemia: mechanisms and management. *Lancet* **ii**, 26–31.

GRUBER K. A. WHITAKER J. M. and BUCKALEW V. M. Jr (1980) Endogenous digitalis-like substance in plasma of volume-expanded dogs. *Nature* **287**, 743–5.

KAYE G. and CAMM A. J. (1985) The role of the atria in fluid volume control. *Brit. J. Hosp. Med.* **34**, 82–8.

EDITORIAL (1980) Dangerous pseudohyponatraemia. *Lancet* **ii**, 1121.

FERRIER I. N. (1985) Water intoxication in patients with psychiatric illness. *Brit. Med. J.* **291**, 1594–6.

LARAGH J. H. (1985) Atrial natriuretic hormone, the renin-angiotensin axis, and blood pressure — electrolyte homeostasis. *New Engl. J. Med.* **313**, 1330–40.

LEE M. R. (1981) The renin-angiotensin system. *Brit. J. Clin. Pharm.* **12**, 605–12.

LEVER A. F., BERETTA-PICCOLI C., BROWN J. J. and DAVIES D. L. (1981) Sodium and potassium in essential hypertension. *Brit. Med. J.* **283**, 463–8.

MACGREGOR G. A. (1983) Nutrition: the changing scene. Dietary sodium and potassium intake and blood pressure. *Lancet* **i**, 750–3.

MAXWELL M. H. and KLEEMAN C. R. (1979) *Clinical Disorders or Fluid and Electrolyte Metabolism.* 3 Ed. McGraw-Hill, New York.

SINGH, S., PADI M. H., BULLARD, H. and FREEMAN H. (1985) Water intoxication in psychiatric patients. *Brit. J. Psychiatry* **146**, 127–31.

WORTH H. G. J. (1983) Plasma sodium concentration: bearer of false prophesies. *Brit. Med. J.* **287**, 567–8.

Potassium

EDITORIAL (1985) Dietary potassium and hypertension. *Lancet* **i**, 1308–9.

EDITORIAL (1983) Adrenaline and potassium — everything in flux. *Lancet* **ii**, 1401–3.

EPSTEIN F. H. and ROAS R. M. (1983) Adrenergic control of serum potassium. *New Engl. J. Med.* **309**, 1450–1.

KASSIRER J. P. and HARRINGTON J. T. (1985) Fending off the potassium pushers. *New Engl. J. Med.* **312**, 785–7.

Calcium, phosphate, magnesium

BRAUNWALD E. (1982) Mechanism of action of calcium channel blocking agents. *New Engl. J. Med.* **307**, 1618–27.

CHERNOW B., SMITH J., RAINEY T. G. and FINTON C. (1982) Hypomagnesaemia; implications for the critical care specialist. *Crit. Care Med.* **10**, 193–6.

DENTON R. M. and McCORMACK J. G. (1981) Calcium ions, hormones and mitochondrial metabolism. *Clin. Sci.* **61**, 135–40b.

EDITORIAL (1981) Diphosphonates: aimed in a chemical sense *Lancet* **ii**, 1326–8.

EDITORIAL (1984) Vitamin D and the lymphomedullary system. *Lancet* **i**, 1105–6.

HOSKING D. J. (1981) Paget's disease of bone. *Brit. Med. J.* **283**, 686–8.

JAMIESON M. J. (1985) Clinical algorithm: Hypercalcaemia. *Brit. Med. J.* **290**, 378–82.

KENNEY J. (1985) Calcium channel blocking agents and the heart. *Brit. Med. J.* **291**, 1150–2.

LEMANN J. and GRAY R. . (1984) Calcitriol, calcium and granulomatous disease. *New Engl. J. Med.* **311**, 1115–7.

LEVINE B. S. and COBURN J. W. (1984) Magnesium the mimic/antagonist of calcium. *New Engl. J. Med.* **310**, 1253–5.

LLINAS R. R. (1982) Calcium in synaptic transmission. *Sci. Am.* **247**, 38–47.

MUNDY G. R., IBBOTSON K. J., D'SOUZA S. M., SIMPSON E. L. *et al.* (1984) The hypercalcaemia of cancer. *New Engl. J. Med.* **310**, 1718–27.

PATERSON C. R. and FEELY J. (1983) Vitamin D metabolites and anologues, diphosphonates, danazol and bromocriptine. *Brit. Med. J.* **286**, 1625–8.

RAISZ L. G. and KREAM B. E. (1983) Regulation of bone formation. *New Engl. J. Med.* **309**, 29–35, 83–9.

RAMBAUT P. C. and GOODE A. W. (1985) Skeletal changes during space flight. *Lancet* **ii**, 1050–2.

STANBURY S. W. (1981) Vitamin D and hyperparathyroidism. *J. R. Coll. Phys.* **15**, 205–17.

SWALES J. D. (1982) Magnesium deficiency and diuretics. *Brit. Med. J.* **285**, 1377–8.

WILKINSON R. (1984) Treatment of hypercalcaemia associated with malignancy. *Brit. Med. J.* **288**, 812–3.

Acid base balance

BISWAS C. K., RAMOS J. M., AGROYANNIS B. and KERR D. N. S. (1982) Blood gas analysis: effect of air bubbles in syringe and delay in estimation. *Brit. Med. J.* **284**, 923–7.

BRYANT M.T.T. (1977) Gases stored in disposable syringes. A study of changes in their concentration. *Anaesthesia* **32**, 784.

COHEN R. D. (1982) Some acid problems *J. R. Coll. Phys.* **16**, 69–77.

HAZARD P. B. and GRIFFIN J. P. (1982) Calculation of sodium bicarbonate requirement in metabolic acidosis. *Am. J. Med. Sci.* **283**, 18–22.

HOOD I. and CAMPBELL E. J. M. (1982) Is pK OK? *New Eng. J. Med.* **306**, 864–5.

JACOBSON H. R. and SELDIN D. W. (1983) On the generation, maintainance and correction of metabolic alkalosis. *Am. J. Physiol.* **245**, (Sept/Oct), F425–32.

MAREN T. M. (1985) Carbonic anhydrase. *New Eng. J. Med.* **313**, 179–81.

ROOS A. and BORON W. F. (1981) Intracellular pH. *Physiol. Rev.* **61**, 296–434.

SIESJO B. K. (1984) Administration of base via the CSF route: a clinically useful; treatment of cerebral acidosis? *Intensive Crit. Care Digest* **3**, 5–9.

SYKES M. K. (1975) The management of acid base balance. In *Anaesthesia Rounds*, Number 9. I.C.I. Ltd.

Fluid balance in the surgical patient

BLAKELY C. and TINKER J. (1983) Vasodilators in acute circulatory failure. *Intensive Care Med.* **9**, 5–11.

CHERNOW B., RAINEY T. G., LAKE C. R. (1982) Endogenous and exogenous catecholamines in critical care medicine. *Crit. Care Med.* **10**, 409–16.

GEYER R. P. (1982) Oxygen transport in vivo by means of perfluorochemical preparations. *New Engl. J. Med.* **307**, 304–6.

JANVRIN S. B., DAVIES G. and GREENHALGH R. M. (1980) Postoperative deep vein thrombosis caused by intravenous fluids during surgery. *Br. J. Surg.* **67**, 690.

LUCAS C. E. and LEDGERWOOD A. M. (1983) The fluid problem in the critically ill. *Surg. Clin. North Am.* **63**, 439–54.

READ D. H. and MANNERS J. M. (1983) Osmolal excretion after open heart surgery. *Anaesthesia* **38**, 1053–61.

Nutrition

CAHILL G. F. (1976) Starvation in man. *Clin. Endocr. Metab.* **5**, 397.

CUTHERBERTSON D. P. (1972) Protein requirements after injury, quality and quantity. In Wilkinson *Parenteral Nutrition*. Churchill Livingtone, Edinburgh.

DAHN M. S. and LANGE P. (1982) Hormonal changes and their influence on metabolism and nutrition in the critically ill. *Intensive Care Med.* **8**, 209–14.

EDITORIAL (1985) Energy and protein requirements revisited. *Lancet* **ii** 1279–80.

LEE H. A. (1974) *Parenteral Nutrition in Acute Medical Illness*. Academic Press, London.

LINDEN M., JEPPSSON R. and EKMAN L. (1986) Catheter occlusion using admixtures for complete IVN in three litre bags. *Brit. J. Parenteral therapy* **7**, 24–6.

MOORE, F. D. (1959) *Metabolic Care of the Surgical Patient*. W. B. Saunders, Philadelphia.

MOORE, R. A., SMITH, R. F. *et al.* (1981) Sex and surgical stress. *Anaesthesia* **36**, 263.

SHERLOCK S. (1984) Nutrition and the alocholic. *Lancet* **i**, 436–9.

WILLATTS S. M. (1986) Nutrition. *Br. J. Anaesth.* **58**, 201–22.

WOOLFSON A. M. J. (1979) Metabolic considerations in nutritional support. *Res. Clin. Forums* **1**, 35–47.

Renal disease

ALFREY A. C. (1984) Aluminium toxicity. *New Engl. J. Med.* **310**, 1113–5.

BERRY A. J. (1981) Respiratory support and renal function. *Anesthesiology* **55**, 655–67.

BEVAN D. R. (1979) *Renal Function in Anaesthesia and Surgery*. Academic Press, London.

BLACK D. and JONES N. F. (1979) *Renal Disease*. 4 Ed. Blackwell Scientific Publications, Oxford.

BLYTHE W. B. (1983) Captopril and renal autoregulation. *New Engl. J. Med.* **308**, 390–1.

BRENNER B. M., MEYER T. W. and HOSTETTER T. H. (1982) Mechanisms of disease. Dietary protein intake and the progressive nature of kidney disease. *New Engl. J. Med.* **307**, 652–9.

CLIVE D. M. and STOFF J. S. (1984) Renal syndromes associated with nonsteroidal antiinflammatory drugs. *New Engl. J. Med.* **310**, 563–72.

CUNDY T. F. and KANIS J. A. (1985) Renal bone disease. *Brit. J. Hosp. Med.* **33** 35–40.

DUSTAN H. P. (1984) Renovascular hypertension and azotaemia. *New Engl. J. Med.* **311**, 1114–5.

EDITORIAL (1982) Ambulatory peritonitis. *Lancet* **i** 1104–5.

EDITORIAL (1983) Haemofiltration. *Lancet* **i**, 1196–7.

EDITORIAL (1984) Who shall be dialysed? *Lancet* **i**, 717.

ESPINEL C. H. and GREGORY A. W. (1980) Differential diagnosis of acute renal failure. *Clin. Nephrol.* **13**(2), 73–9.

HABER E. (1984) Renin inhibitors. *New Engl. J. Med.* **311**, 1631–3.

KNAPP M. S. (1982) Renal failure — dilemmas and developments. *Brit. Med. J.* **284**, 847–50.

Luke R. G. (1983) Renal replacement therapy. *New Engl. J. Med.* **308**, 1593–5.

Specific problems

BRASS E. P. and THOMPSON W. L. (1982) Drug induced electrolyte abnormalities. *Drugs* **24**, 207–28.

CALMAN K. C. (1982) Cancer cachexia. *Brit. J. Hosp. Med.* **27**, 28–34.

COOMBES R. C. (1982) Metabolic manifestations of cancer. *Brit. J. Hosp. Med.* **27**, 21–7.

EDITORIAL (1982) Management of heat stroke. *Lancet* ii, 910–11.

EDITORIAL (1982) Hormones and fluid retention in cirrhosis. *Lancet* i, 1341–2

EDITORIAL (1984) Hepatic encephalopathy today. *Lancet* i, 489–91.

FOSTER D. W. and McGARRY J. D. (1983) The metabolic derangements and treatment of diabetic ketoacidosis. *New Engl. J. Med.* **309**, 159–69.

GORDON I. J., BOWLER C. S. COAKLEY J. and SMITH P. (1984) Algorithm for modified alkaline diuresis in salicylate poisoning. *Brit. Med. J.* **289**, 1039–40 and **290** 155 (1985).

HAYES P. C., STEWART W. W. and BOUCHIER I. A. D. (1984) Influence of propranolol on weight and sodium and water homeostasis in chronic liver disease. *Lancet* ii, 1064–8.

LEWIS G. B. H. and HECKER J. F. (1985) Infusion thrombophlebitis. *Brit. J. Anaesth.* **57**, 220–33.

McCLOY R. F. (1984) Metabolic problems associated with gastrointestinal and pancreatic disease. *Brit. J. Anaesth.* **56**, 83–94.

MUIR I. F. K. and BARCLAY R. L. (1974) *Burns and their Treatment.* 2nd Ed. Lloyd-Luke, London.

PARISH P. (1982) Benefits to risks of intravenous therapies. *Brit. J. Intravenous Therapy* **3**, 10–19.

PRESCOTT L. F. (1983) New approaches to managing drug overdose and poisoning. *Brit. Med. J.* **287**, 274–6.

SETTLE J. A. D. (1974) Urine output in burns. *Burns* **1**(1), 23–42.

SYMPOSIUM (1982) Modern trends in burns care. *J. R. Soc. Med.* **75**, 1–51 (Supplement).

TUCKER R. M., VAN DEN BERG C. J. and KNOX F. G. (1980) Diuretics: role of sodium balance. *Mayo Clin. Proc.* **55**, 261–6.

WATKINS P. (1982) Diabetic emergencies. *Brit. Med. J.* **285**, 360–3.

WHARTON B. A. (1981) Gastroenteritis in Britain: management at home. *Brit. Med. J.* **283**, 1277–8.

WINEGRAD A. I., KERN E. F. O. and SIMMONS D. A. (1985) Cerebral oedema in diabetic ketoacidosis. *New Engl. J. Med.* **312**, 1184–5.

Practicalities of setting up infusion lines

BRECKENRIDGE A. and WOOLFSON A. M. J. (1984) Drugs and intravenous fluids. *Brit. Med. J.* **289**, 637–8.

GEORGE R. J. D. (1980) Practical procedures. How to insert a flotation catheter. *Br. J. Hosp. Med.* **23**(3), 296.

LOWE G. D. (1981) Filtration in i.v. therapy. *Brit. J. Intravenous Therapy* **2**, 37–54 (June), 24–38 (Dec).

PETERS J. L. (1982) Current problems in central venous catheter systems. *Intensive Care Med.* **8**, 205–8.

PETERS J. L. FISHER C. and MEHTA S. (1980) Intravenous stopcocks and injection ports. *Lancet* ii, 701.

WILLATTS S. (1980) Practical procedures. How to cannulate a peripheral artery. *Br. J. Hosp. Med.* **23**(6), 628.

Index